Peter Hood

How Time is Measured

SECOND EDITION

Oxford University Press
1969/-

This 24-hour dial of the great clock at Hampton Court Palace is believed to have been made in 1540 for Henry VIII. Noon is at the foot of the dial. The hours are painted on the stonework, and within this circle of figures are 3 copper disks, turning at different speeds. The centre disk, with the globe of the earth, is divided to show the 4 quarters of the moon, and a hole with a moving shutter shows the phases. White figures show when the moon crosses the meridian. The second disk gives the moon's age in days, and to this is fastened the hour hand. The largest disk, nearly 8 feet diameter, shows the days and months, and the signs of the zodiac divided in degrees. These are read from the hour hand, the symbol of the sun showing its position in the zodiac.

Contents

The publishers gratefully acknowledge the help given in the production of this book by the late Sir Harold Spencer Jones, F.R.S.; and that of Mr. Humphrey M. Smith, B.Sc., M.I.E.E., with articles 26, 27, and 28 of this revised edition. Photographs and drawings, other than those acknowledged are by the author or his brother, Cleveland Hood.

1 The Origin of the Calendar

WHEN people first started to record time, the most obvious event was the rising and setting of the sun, caused by the daily rotation of the earth. For measuring longer periods, they used the phases of the moon, which recur month by month in the same order; and by noting the time of each new moon they were able to compile a calendar. But the lunar or moon-month does not consist of an exact number of days (it may be 29 or 30), and 12 such months do not make a complete year. A calendar of this sort would make each year begin a little earlier than the year before, and it would soon become obvious that the calendar was getting out of step with the seasons. To remedy this, an extra month, called an intercalary month, was slipped, now and again, into the calendar. There still exists, for example, a letter from Hammurabi, King of Babylon about 1900 B.C., ordering one of his governors to add a month to the calendar in this way.

The ancient Egyptians also used a lunar calendar and the year was given, from time to time, a thirteenth month, to keep it in approximately correct relationship to the seasons. While this lunar calendar continued in use for religious purposes, the Egyptians later made a more accurate calendar for civil purposes. They adopted 12 months of 30 days, with 5 feast days added at the end of the year, making 365 days. But the year cannot be divided into an exact number of days; it is in fact nearly a quarter of a day longer. So after this civil calendar had been in use for some centuries, it was found that it was gradually drifting with respect to the seasons.

It was Julius Caesar who sought a more reliable basis for the calendar. He consulted the astronomer Sosigenes of Alexandria, and the outcome was the idea of making every fourth year a leap year with 366 days instead of 365 days. This, usually called the Julian calendar, and leaving much to be desired, was adopted in the year 45 B.C.; and at the same time the Romans changed their calendar year so that it began in January, instead of in March as was the previous custom. The Romans did not number the days of their months consecutively. Three dates were specially named: the Kalends or first day of the month; the Nones, which might be the 5th or the 7th of the month; and the Ides, nominally the day of the full moon, but fixed at the 13th or 15th. Any intervening day was said to be so many days before the next named day; though both days were included in the counting.

The names of the months we use are those used by the Romans. The Latin months were originally Martius, Aprilis, Maius, Iunius, Quintilis, Sextilis, September (seventh month), October, November, December, Ianuarius, Februarius. The summer month Quintilis was renamed Julius by Caesar; and Augustus who followed him gave his own name to the next month. There have been no changes since, and the lengths of the months now in use remain exactly as they were fixed by Julius Caesar.

The names of the week-days are mainly Anglo-Saxon, but they have a much older history. The seven-day week is unlike the day, month, or year in that it does not correspond to any celestial period. It had its origin in astrology, the seven days being named after the seven 'planets' or their divinities, as they were known to the ancients. This seven-day week was taken over by the Jews from the Assyrians, and then from the Jews by the Christians. So we have Sun's day, Moon's day; followed by the days of Mars, Mercury, Jupiter (or Jove), and Venus. These, in French, became Mardi, Mercredi, Jeudi, Vendredi; but in Anglo-Saxon they were named after the equivalent heathen gods, Tiw's day, Woden's day, Thor's day, and a heathen goddess, Figg's day; the seventh day being Saturn's day.

The Christian era began with the birth of Christ, and so we use B.C. for the dates 'before Christ', and A.D. or Anno Domini, 'in the year of our Lord', for dates subsequently. The method was first introduced into England by the Anglo-Saxon historian, Bede. It is believed that the original reckoning is a little in error, and that the actual birth of Christ may have been in 4 B.C.

Sometimes dates are given in Roman numerals. The original practice was to repeat the figures as often as required, like IIII on a clock dial. In later usage, a lesser figure before a greater one is always subtracted from it: MCMLIX = 1959, and MCMLX = 1960.

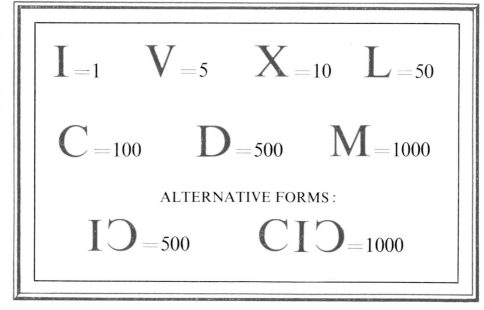

2 Perfecting the Calendar

A YEAR is the time it takes the earth to make one complete revolution round the sun. There are different ways of measuring this, and as the calendar must be made to fit the seasons, the year is measured from one March equinox to the next. This is known as the tropical year, and is not the same as the year measured by the stars. The Julian calendar was based on a year of 365¼ days, but the length of the tropical year is slightly less, to be exact, 365 days 5 hrs. 48 min. 46 sec. Consequently, over many centuries, there was an accumulating error, so that the Easter celebrations were getting too far from the spring equinox. Pope Gregory XIII decided that the error, which then amounted to 10 days, must be corrected. With the advice of the astronomers Clavius and Lilius it was ordered that the day following 4 October 1582 was to become 15 October. It was also arranged that century years should not be leap years unless divisible by the figure 400.

This, in effect, means a slight reduction in the number of leap years; and though this Gregorian calendar, as now used, is not absolutely accurate, nearly 4000 years will have passed before it becomes a day in error. At the time of the change in the sixteenth century, Britain, having broken from the Roman Catholic Church, was in no mood to follow an edict of the Pope; so the revised calendar was not adopted in England until 1752; by this time the Julian calendar, which the English were still using, was 11 days behind. To put this right it was decreed that the day after 2 September should be called 14 September. This much disturbed some people, who thought that their lives had been made shorter or their pay deducted, and they cried 'Give us back our eleven days'. At the same time another change was made: the Church in the Middle Ages had adopted the practice of beginning the legal year on 25 March, and this was now altered to 1 January. But the British government did not change its ways, and to the present day the financial year is still reckoned from 6 April, the new calendar equivalent of 25 March (11 days difference).

Our calendar is far from perfect. The disadvantage of having Christmas day fall on any day of the week, and Easter day any time between 22 March and

January					
Sun	1	8	15	22	29
Mon	2	9	16	23	30
Tue	3	10	17	24	31
Wed	4	11	18	25	·
Thu	5	12	19	26	·
Fri	6	13	20	27	·
Sat	7	14	21	28	·

February					
Sun	·	5	12	19	26
Mon	·	6	13	20	27
Tue	·	7	14	21	28
Wed	1	8	15	22	29
Thu	2	9	16	23	30
Fri	3	10	17	24	·
Sat	4	11	18	25	·

March					
Sun	·	3	10	17	24
Mon	·	4	11	18	25
Tue	·	5	12	19	26
Wed	·	6	13	20	27
Thu	·	7	14	21	28
Fri	1	8	15	22	29
Sat	2	9	16	23	30

April					
Sun	1	8	15	22	29
Mon	2	9	16	23	30
Tue	3	10	17	24	31
Wed	4	11	18	25	·
Thu	5	12	19	26	·
Fri	6	13	20	27	·
Sat	7	14	21	28	·

May					
Sun	·	5	12	19	26
Mon	·	6	13	20	27
Tue	·	7	14	21	28
Wed	1	8	15	22	29
Thu	2	9	16	23	30
Fri	3	10	17	24	·
Sat	4	11	18	25	·

June					
Sun	·	3	10	17	24
Mon	·	4	11	18	25
Tue	·	5	12	19	26
Wed	·	6	13	20	27
Thu	·	7	14	21	28
Fri	1	8	15	22	29
Sat	2	9	16	23	30

L

25 April is well known. The fixing of Easter has always been a problem for the Church. As now celebrated, Easter day is the first Sunday after the full moon occurring on or next after 21 March. Tables making these calculations are to be found at the beginning of the Book of Common Prayer. There are other considerable inconveniences in the present-day calendar. The variations in the lengths of the different months, which have been continued since Roman times, are unreasonable. Any given date falls on different days of the week in successive years, and with much trouble we have to have a fresh calendar compiled each year. The number of pay-days in a year may vary, and the four quarters of the year are of unequal length.

A simple calendar, known as the World Calendar, has been proposed. It divides the year into four identical quarters, the first month of each quarter beginning on a Sunday and having 31 days, the second beginning on a Wednesday and having 30 days, the third beginning on a Friday and having 30 days. Every month thus has 26 week-days, whereas in our present calendar it may have 24, 25, 26, or 27. Each quarter contains 91 days, precisely 13 weeks. The four quarters make a total of 364 days. In order to bring it up to 365 an extra day as a universal holiday, and not counted as part of the week, is added at the end of each year. In leap years only, another extra day would also be added at the end of June. These extra days would have the dates of 31 December and 31 June; but being outside the week would not be named by any week-day but would have their own names, World day and Leap day; after which the week-days would continue where they last left off.

When previous reforms of the calendar occurred, it was only a matter of changing a date, and the sequence of the days of the week was not disturbed. People are reluctant to change age-old habits; but, in fact, the idea of having an extra day during the week is not really entirely new. Already we have to add a day, or drop a day, when crossing the international Date Line, as described later in this book.

July					
Sun	1	8	15	22	29
Mon	2	9	16	23	30
Tue	3	10	17	24	31
Wed	4	11	18	25	·
Thu	5	12	19	26	·
Fri	6	13	20	27	·
Sat	7	14	21	28	·

August					
Sun	·	5	12	19	26
Mon	·	6	13	20	27
Tue	·	7	14	21	28
Wed	1	8	15	22	29
Thu	2	9	16	23	30
Fri	3	10	17	24	·
Sat	4	11	18	25	·

September					
Sun	·	3	10	17	24
Mon	·	4	11	18	25
Tue	·	5	12	19	26
Wed	·	6	13	20	27
Thu	·	7	14	21	28
Fri	1	8	15	22	29
Sat	2	9	16	23	30

October					
Sun	1	8	15	22	29
Mon	2	9	16	23	30
Tue	3	10	17	24	31
Wed	4	11	18	25	·
Thu	5	12	19	26	·
Fri	6	13	20	27	·
Sat	7	14	21	28	·

November					
Sun	·	5	12	19	26
Mon	·	6	13	20	27
Tue	·	7	14	21	28
Wed	1	8	15	22	29
Thu	2	9	16	23	30
Fri	3	10	17	24	·
Sat	4	11	18	25	·

December					
Sun	·	3	10	17	24
Mon	·	4	11	18	25
Tue	·	5	12	19	26
Wed	·	6	13	20	27
Thu	·	7	14	21	28
Fri	1	8	15	22	29
Sat	2	9	16	23	30

3 The Definition of Time

THE regular succession of day and night is due to the rotation of the earth on its axis. The rotation is towards the east, so when our part of the earth's surface moves into the sunshine, the sun appears to rise in the east. After the period of daylight, it appears to set in the west, because we are carried round into the night or shadowed side of the earth. This sequence is continually repeated. As civil time-keeping is based on the length of the day—the hours of daylight and darkness combined—such a measurement might seem quite simple, since it is only a matter of finding out how long it takes the earth to make one complete rotation. This is not really so easy to measure. The most accurate way of finding out is by observing the stars, since the stars, like the sun, appear to rise and set owing to the earth's rotation.

Observation may be made with a transit instrument, which is a telescope set in trunnions so that it may swing only in the meridian, or north to south line, and cannot turn sideways. If we note the instant when a bright star crosses the meridian, and the next night note the instant when the same star again crosses the meridian, the intervening time represents a complete day of 24 hours. Time determined from observations of star transits is called sidereal time, after *sidus* the Latin for a star. Astronomers use sidereal time, but it is no use to anyone else, because it does not coincide with solar time, measured by the sun. The length of the day from noon, when the sun crosses the meridian, to the next noon, is some 4 minutes longer than the day measured by the stars. The diagram at the foot of the page opposite shows the reason for this difference between solar and sidereal time.

Ordinary solar time cannot be used for clocks for it is not quite uniform, and the length of the solar day varies slightly in the course of the year. The

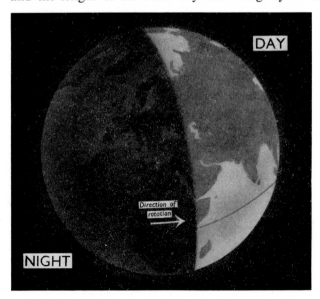

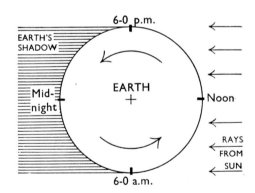

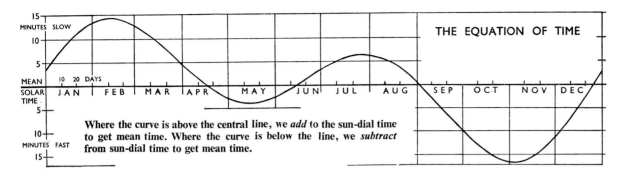

THE EQUATION OF TIME

Where the curve is above the central line, we *add* to the sun-dial time to get mean time. Where the curve is below the line, we *subtract* from sun-dial time to get mean time.

explanation is rather complicated: the earth moves faster in its orbit in January when nearest to the sun, than in July when furthest from the sun; and also, because of the inclination of the earth's axis, the sun's apparent movement relative to the equator is not uniform. To overcome these difficulties, astronomers introduced the concept of an imaginary sun (called the fictitious mean sun) which moves at a constant rate throughout the year. This gives us mean solar time, and the difference between this mean solar time (as used for clocks) and apparent solar time (as seen on a sun-dial) is called the equation of time; the word 'equation' being used in its old sense of 'correction'. It varies throughout the year, and may sometimes amount to as much as 16 minutes. There are only four dates when mean and solar time are the same. Therefore on all other dates, when reading time on a sun-dial, we have to add or subtract a few minutes, according to the scale shown in the diagram above, to get mean time.

Since the times of sunrise and sunset are variable, we use noon as the fixed event from which the hours are reckoned. Each day starts at midnight with the 12 hours up to noon, these hours being marked a.m. (for the Latin *ante meridiem*, 'before noon'); and the 12 hours following noon are marked p.m. (for *post meridiem*, 'after noon'). As it is often confusing to have to use a.m. and p.m., the 24-hour clock is now coming into more general use for official purposes. Beginning at midnight, the hours are counted 0 to 23. So the hours a.m. are the same as on an ordinary clock; but 1.0 p.m. becomes 13 hrs., 6.0 p.m. becomes 18 hrs., and so on. For many years the 24-hour system has been in common use on the Continent, especially for railway and airline tables, and it is now much used in Great Britain. We write 16.35 for what used to be 4.35 p.m.

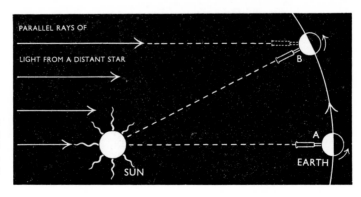

The different lengths of the day measured by the sun or by a star. A fixed telescope at *A*, pointing to a star, will again point to the star after the earth has turned once. In the case of the sun, rather more than one rotation is required, as at *B*, because the earth has moved on in its orbit.

9

4 Sun-dials

SUN-DIALS were used by the ancient Egyptians, and by the Greeks and Romans. They were the earliest method of telling the time. Several varieties were used, and when correctly made they could be quite accurate. In England, the earliest sun-dials still standing date from Anglo-Saxon times. They are crudely designed, but no one had then any accurate means of telling the time; and even King Alfred burned candles to know how the day progressed, each candle lasting four hours.

The long picture shows a sun-dial which dates from a few years before the Norman conquest. A rod or 'gnomon' originally projected from the centre hole, and its shadow fell across the dial when the sun shone; but as the gnomon probably stuck out at right angles to the wall, the readings would only be correct at sunrise, noon, and sunset. The Saxons divided their day into four 'tides', and we still have the words noontide and eventide. On this dial these have been subdivided to make eight divisions; and the early morning one, which was probably a time of service, has an X on it. During the Middle Ages, churches often had sun-dials scratched roughly on a wall.

In the sixteenth century, when watches were expensive rarities, portable or pocket sun-dials became very popular. Those which showed the time from the sun's direction were fitted with a compass so that they could be placed fairly correctly. The simplest dials showed the time by the sun's altitude, and a common form was the ring dial. We read in *As You Like It*, 'And then he drew a dial from his poke', and probably it was one of these. They were in use a long time, and some were still being made at Sheffield up to the end of the eighteenth century. The dial was held by its suspension ring. There was a small hole in the sliding part on the outside, and this was turned to the right position for the time of the year, so that the sun's rays passing through the hole made a bright spot of light on the inside of the ring, where there was a scale of hours.

Even after clocks came into general use, sun-dials of one sort or another were still used because clocks were inaccurate and had to be re-set frequently, and there were no time signals or other means of getting the time. On the facing page is a typical garden sundial of the eighteenth century. The ornamented part is merely the support for the sloping gnomon, the shadow of which falls on the horizontal plate

SUN'S RAYS

Pocket ring sun-dial, dated 1716.

[Crown copyright reserved]

10

which is engraved with the numbers and divisions of the hours. With a sun-dial we cannot usually read the dial to an accuracy much greater than about a couple of minutes. This is because the sun is not a point of light which would give a sharp shadow, but is a visible disk of half a degree in diameter, and this makes the shadow slightly diffuse.

The angle of the gnomon with the horizontal plate has to be the same as the latitude of the place where the dial is fixed: for London this is $51\frac{1}{2}°$. The gnomon points upwards to the celestial pole and must be set exactly north and south, that is, in the meridian; and it must point true north, not magnetic north, which is slightly different, as will be seen from a map. When the dial is fixed to a vertical wall, the gnomon must be set at an angle with the wall which is 90° minus the latitude of the place: for London, $38\frac{1}{2}°$. The wall must face due south, and if it does not, a more elaborate geometrical construction may be necessary to mark the dial correctly and place the gnomon so that its support does not hide the hours.

Once a gnomon is set in the meridian, and at the correct angle, we may, of course, make our own dial by marking each hour by the aid of a watch. But we have to remember that the time shown by a sun-dial is the local solar time; it is not mean time. So we must allow for the difference, according to the equation of time, when converting mean time to solar time, or solar time to mean time. Also we have to make an allowance for longitude if we are not on the line of longitude of the mean time we are using. In Britain, this is longitude 0°, and for every degree of longitude east of this the solar time is 4 minutes in advance; or for every degree west, 4 minutes behind, as described on later pages. This explains why people are sometimes puzzled to find a sun-dial not showing the same time as a watch.

A garden sun-dial, dated 1718.

5 Early Time-keeping

WATER clocks were sometimes used in ancient times as well as sun-dials. In Egypt, a water clock was a large bowl with sloping sides and a tiny hole at the bottom. When the bowl was filled, the water slowly ran out; and as the level of the water went down, marks on the inside of the bowl showed the hours. Another kind, which became common in Roman times, was a small bowl which was floated in a bucket of water. The bowl had a tiny hole in it, so that it gradually filled with water and sank after a fixed interval of time. A water clock called a clepsydra, used by the Greeks and Romans, was more complicated. Water flowed slowly into a cylinder or tank in which was a float carrying an arm or pointer which moved over a scale of hours.

Astronomers have always divided the day into 24 equal hours; but from very early times and even up to the Middle Ages it was customary, for ordinary use, to divide the daylight into 12 hours, and the night into 12 hours. Some people began the day at sunrise, others at sunset. Thus in the account of the Creation, in the 1st chapter of Genesis, we read, 'The evening and the morning were the first day'. The daylight period thus consisted in winter of 12 short hours and in summer of 12 long hours. These variable hours were known as temporal hours and, as they were reckoned from sunset and from sunrise, the numbering was quite different from that of today.

We read in St. Luke, 'And it was about the sixth hour, and there was a darkness over all the earth until the ninth hour. And the sun was darkened. . . .' Now when the numbering began at what we call 6.0 a.m., the sixth hour would actually be noon, and the ninth hour was, therefore, 3.0 p.m. This method of reckoning persisted in the Middle Ages when the periods of time were known as the canonical hours, being named after the times of prayer: Matins, Lauds, Prime, Terce, Sext, Nones, Vespers. The Nones were later shifted to midday, giving us the word noon. With the introduction of mechanical clocks, about the fourteenth century, the division of the day into 24 equal hours gradually came into general use. To distinguish between time on this system and on the temporal hour system, it became customary to refer to the former as 'of the clock', which still survives in our 'o'clock'. The numbering of the hours from 1 to 12 twice over instead of from 1 to 24 has been carried over from the temporal hour system.

Used in Roman Britain as a water clock, this thin metal bowl has a hole in the centre. Below is an old sand glass, about 1700 A.D.

The sand-glass or hour-glass was invented in the Middle Ages and, while clocks were still not easily available, continued to be used up to the early nineteenth century for measuring short periods of time, such as the length of a speech or a sermon. When all the sand had run into the lower bulb, the glass was merely reversed when next used. Sand-glasses were used on board ship for timing the periods of watch, and for timing the ship's speed, when the log was run out and the number of knots on the line passing in a given time were counted.

THE first mechanical clock was probably no more than an alarm mechanism which indicated certain hours, so that the keeper of the clock could himself strike a bell. Subsequently the striking was done automatically, sometimes with little figures called jacks, which moved and hit the bells. In the time of Edward III a clock tower was built in the courtyard opposite the entrance to Westminster Hall, and from 1371 onwards there are various accounts of the cost of repairs and upkeep of the clock, as well as the payment for the keeper of the 'King's clock in the palace of Westminster'. The old clock tower was allowed to become ruinous towards the end of the seventeenth century and was not replaced until the present tower of Big Ben was built in Victorian times.

The great clock at Rouen in Normandy was set up in 1389 and, though parts of this famous clock have been altered or replaced, this same clock which tolled the hours in the long-drawn-out trial of Joan of Arc is still working. Some of the early clocks in England survive as curiosities; that of Salisbury, believed to date from 1386, is the oldest; though the Wells cathedral clock, whose original movement or mechanism is preserved in the Science Museum, London, is probably as old, and continues to strike the hours and quarter-hours.

[Phillips' City Studio]

'Jack Blandifer' in Wells Cathedral, who strikes the hour bell with a hammer, and the quarter bells with his heels.

When a clockwork toy is wound up, and the wheels set going, they whirr round until the spring runs down. With a clock that has to show the time, there must be some method of allowing the wheels to turn slowly at a pre-arranged rate. This is done by a wheel, called the escape wheel, which is released only a tooth at a time. The mechanism which does this is called the escapement, and the care with which it is made and adjusted decides the accuracy of the time-keeping. The first diagram overleaf shows the simple verge escapement, as used in the oldest clocks. Though weights on the cross-bar or balance arm could be moved inwards or outwards to alter the speed, this mechanism was never accurate because the speed also depended on the power of the drive, which was apt to vary. For small clocks and watches, a balance wheel took the place of the cross-bar, the wheel swinging to and fro in the same way.

The movement of a striking clock is in two parts, and in early clocks each part was driven by a weight suspended by a rope wound round a wooden drum. The first part, with the gears driving the escapement, is called the time train; and the second part, the train of gears driving the striking mechanism, is called the striking train, and is shown in the next diagram. The striking must not only be released each hour: the right number of hours must be struck. This is done by the peculiar wheel, called a locking plate, driven by gear teeth on the inside. Its smooth outer rim has notches at different distances, into which a lever drops. It is this lever which is raised to start the striking train. The shaft which spins fastest has on it a two-armed fan called a fly to prevent the mechanism from going too fast. When the required number of hours have been struck, the lever drops into a notch in the right place on the locking plate, and the striking train is stopped until released again at the next hour.

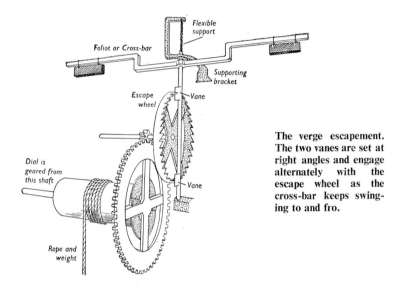

The verge escapement. The two vanes are set at right angles and engage alternately with the escape wheel as the cross-bar keeps swinging to and fro.

The early iron clocks, being large and heavy, were made by blacksmiths. The first household clocks were of a rather similar pattern, though on a much smaller scale. They had to be more accurately constructed, with the reduction in size, and they were generally the work of locksmiths or gunsmiths. Clockmaking began to be a craft when, in the time of Elizabeth I, clocks were made throughout of brass. In the Middle Ages there had been no threaded screws or nuts and bolts, both of which are necessary in the making of smaller clocks; so that clockmakers who began to use them were the forerunners of modern mechanical engineers. By the seventeenth century the characteristic English domestic clock was the lantern clock, sometimes called a Cromwellian. In the early examples there was only one hand, the hour hand, and this pointed to an inner ring with the hour divided into quarters. As the verge escapement was still used, such a clock was never accurate and might be a quarter of an hour fast

A lantern clock, made by Thomas Knifton, Lothbury, London, about 1650.

or slow. By inventions later in the century, time-keeping was so much improved that it was found possible to add long minute hands to clocks, pointing

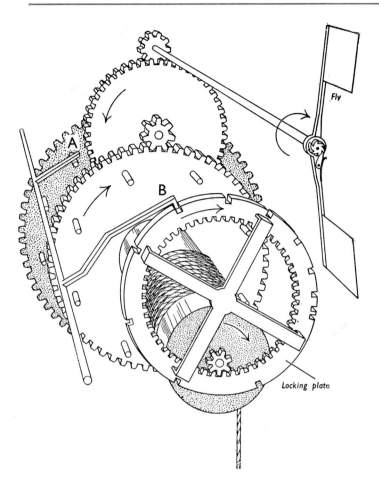

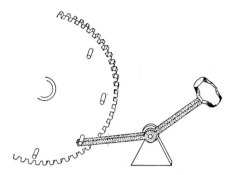

Pins projecting from a wheel of the time train (in background) lift lever *A* each hour, and so *B* is raised from a notch in the locking plate, thus releasing the striking train. Above : how the hammer lever is worked by pins projecting from the great wheel of the striking train.

to minutes engraved on an outer ring. Nowadays the inner ring is generally left out, though the hour hand still points to where the inner ring used to be.

The pendulum came into general use for clocks in the time of Charles II. At first it was used with the old verge escapement, in place of the cross-bar, but the invention of the anchor escapement, described on the next pages, soon followed. Also, about the same time, came the invention of the hair spring to control the small balance wheel in watches. These changes led to a great improvement in time-keeping. The old verge escapement with its cross-bar, or with a balance wheel, was unreliable because it had no fixed period of swing or oscillation. But both the pendulum and the spring-controlled balance wheel have a natural period of oscillation. As this period is accurately known it serves as the principle on which most of our modern time-keeping is based.

The best clockmakers achieved some precision in their art, and English clocks became world-renowned. The most famous of the London clockmakers, partners in the same firm, were Thomas Tompion (1637–1713) and George Graham (1673–1751). They became Fellows of the Royal Society, and as a mark of honour both were buried in Westminster Abbey. The use of a long pendulum, which beat seconds, led to the popularity of long case clocks, now known as grandfather clocks, from the late seventeenth to the early nineteenth century. The case was generally made by a cabinet maker and decorated according to the fashion of the time. There was often a small aperture cut in the dial in which appeared each day the date of the month; and sometimes above the dial were other devices, such as an indication of the age of the moon.

7 Pendulum Clocks

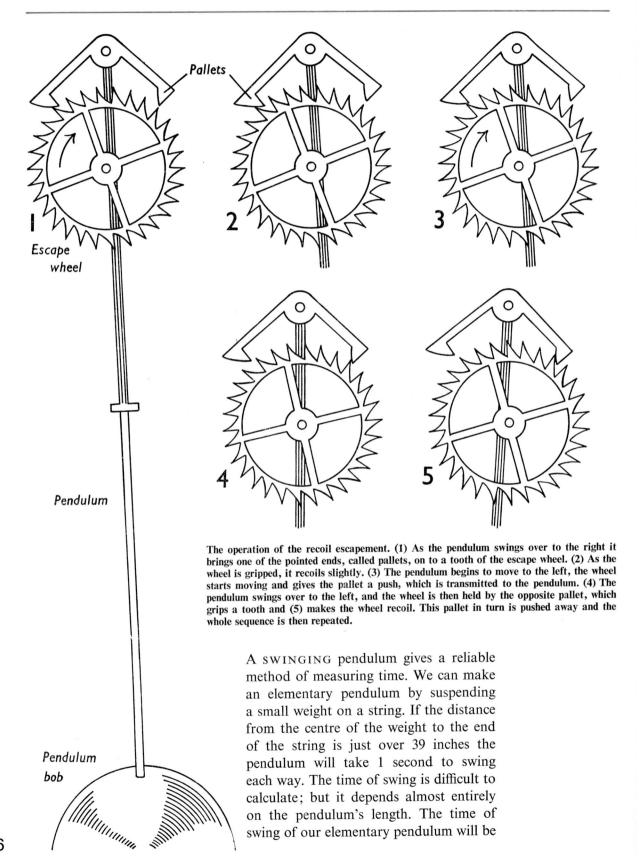

Pallets

Escape
wheel

1

2

3

Pendulum

4

5

The operation of the recoil escapement. (1) As the pendulum swings over to the right it brings one of the pointed ends, called pallets, on to a tooth of the escape wheel. (2) As the wheel is gripped, it recoils slightly. (3) The pendulum begins to move to the left, the wheel starts moving and gives the pallet a push, which is transmitted to the pendulum. (4) The pendulum swings over to the left, and the wheel is then held by the opposite pallet, which grips a tooth and (5) makes the wheel recoil. This pallet in turn is pushed away and the whole sequence is then repeated.

Pendulum
bob

A SWINGING pendulum gives a reliable method of measuring time. We can make an elementary pendulum by suspending a small weight on a string. If the distance from the centre of the weight to the end of the string is just over 39 inches the pendulum will take 1 second to swing each way. The time of swing is difficult to calculate; but it depends almost entirely on the pendulum's length. The time of swing of our elementary pendulum will be halved if its length is quartered, to $9\frac{3}{4}$ inches. The time of swing does vary slightly with the size of the swing; but for small swings the variation may be ignored. This is why the swings of the pendulum in a clock are kept small.

If a pendulum is made of a single metal, its length and therefore its time of swing change with changes in temperature. In domestic clocks this is not important; but in accurate work it must be avoided. One way is to make the pendulum of two metals which expand different amounts with rises of temperature. They are arranged so that as one, usually steel, expands downwards the other expands more, upwards, leaving the effective length of the pendulum unchanged. The second metal may be either brass, in the form of rods, or mercury in a glass cylinder. More simply, the pendulum may be made of a special alloy of nickel and iron, called invar, which hardly changes its length with temperature.

A pendulum clock, driven by a weight or a spring, counts the number of swings and shows the answer on the clock face. The escapement that does the counting imparts a gentle push to the pendulum to keep it going. To adjust the time-keeping, the bob may be screwed up or down. Lowering it makes the clock go slower; raising it, faster.

The diagrams show the operation of a typical anchor, or recoil, escapement: 'anchor' from its shape; 'recoil' from the momentary reversal of the escape wheel that occurs towards the end of each swing. If there is a seconds hand we can readily observe the effect. The sequence shown is continually repeated and it is the pallets that give the familiar tick tock. It requires two swings of the pendulum to release the wheel a distance of one tooth. The shape of these escape wheel teeth, and of the pallets, is improved in more modern designs so that there is no recoil, but the general principle is the same. To allow the pendulum to swing as freely as possible, the anchor carrying the pallets is not directly connected to the pendulum but is pivoted on a separate shaft. A fork-ended lever imparts the motion to the pendulum. The diagram shows how the pendulum is not suspended from a pivot of any kind, but by a very flexible thin steel strip.

A domestic clock should not gain or lose more than about 10 seconds a day. We describe the *error* of a clock as the difference between indicated time and actual time. We state it as the amount to add to the indicated time to give true time; so that an error of $+10$ seconds means that the clock is 10 seconds slow. The *rate* of a clock is the accumulated error over a given period, usually 24 hours.

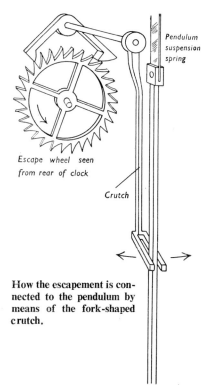

Escape wheel seen from rear of clock

Pendulum suspension spring

Crutch

How the escapement is connected to the pendulum by means of the fork-shaped crutch.

8 The Striking Clock

THE SMALL domestic clock which strikes the hours has been popular for hundreds of years. In the earlier examples the striking was operated by the locking plate mechanism already described; but if the hands are altered, the time becomes out of step with the striking. To overcome this disadvantage, the rack method of striking, in which the striking depends entirely on the position of the hands, was invented in 1676 and has since been used for domestic clocks.

On these pages are illustrated the movement of a modern striking clock which strikes the hours and also gives one stroke at each half-hour. There are two springs to drive the movement; each is set in a drum, called a barrel, on which is the main or great wheel, which does the driving. One end of the coiled steel spring is fastened to the central arbor or axle, the other end to the inside of the barrel. In this way, when the spring is wound, it continues to drive the barrel until it has run down. The left-hand barrel drives the striking train, and the levers controlling the striking mechanism are seen on the front of the movement. The right-hand barrel drives the time train, and this train of gear wheels has to be designed so that the centre wheel, which drives the hands, turns once each hour.

The diagram shows the method by which the hands are driven. The centre wheel is not fixed to the centre shaft, but is pressed against the broader part of the shaft by a friction spring *C*, having four radial arms. This means that the centre wheel normally turns with the shaft being held to it. But if anyone alters the hands of the clock, the centre shaft turns, and the spring arms of *C* slip round the centre wheel without actually disturbing it or the time train of the clock. On the end of the centre shaft is fixed the minute hand, the fitting being of square section, so that the hand cannot slip. Also on this shaft is the trip arm *D*, which releases the striking, and a pinion, *E* in diagram. The hour hand must turn once for every 12 revolutions of the minute hand, so there is a reduction gearing in clocks and watches known as the motion work, the gears here being marked *E*, *F*, *G*, *H*. A short tube

The movement of a striking clock, showing the front, with dial removed.

[Clock by courtesy of The Garrard Engineering and Manufacturing Co. Ltd.]

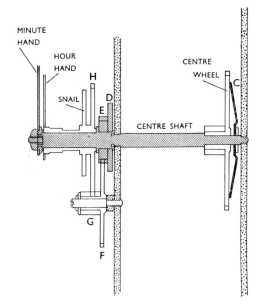

In the diagram, the centre shaft, and parts turning with it, are shown in red. Above: a mainspring in its barrel. On right: the pendulum and its connections.

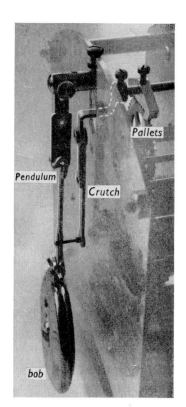

turns freely on the centre shaft, and fixed to this tube is the hour hand, the gear wheel *H*, known as the hour wheel, and a peculiar cam called a snail, because of its shape, the purpose of which is to control the striking.

At the rear of the clock is the pendulum, and also the arm of the hammer which strikes the gong. The pendulum bob may be screwed up or down by a regulating nut, $\frac{1}{4}$ turn making a difference of about 1 minute a day. The pendulum is 3·8 in. long, and it beats $3\frac{1}{6}$ times a second. It will be noticed that it is not a solid rod: the bob is hooked on. There is another joint further up, the purpose of which is to avoid strain on the suspension spring at the top in case the clock, when lifted, is given a forward or backward tilt. A sideways tilt cannot harm the suspension or the pallets in contact with the escape wheel because the crutch, holding the pendulum, passes through a slot which limits its movement. The crutch is made in two parts so as to be adjustable, merely by pressure with the fingers, if the clock needs setting to beat evenly each way.

The striking train has, of course, no escapement; but the small fly or fan that whirrs round prevents the movement going too fast. In the gear train is a star wheel, and as it revolves, each point in turn lifts a projecting pin attached near the pivot of the hammer, thus repeatedly raising the hammer and letting it fall. To see the mechanism which controls the striking we must look once more at the front of the clock movement. When the dial is removed, ordinary clocks

19

are found to have merely the motion work under the hands. In a striking clock, however, not only is the snail attached to the hour wheel, but there is also a rack, with teeth like a segment of a gear wheel, pivoted on a long arm so that when released, the end or tail swings over and makes contact with the edge of the snail. The shape of the snail decides how many strokes of the hammer there shall be. If twelve are needed, the tail of the rack drops a long way; if only one, the snail is broad, and so the tail moves only a little way.

In order to understand exactly how the striking is controlled we must follow very closely the whole sequence of the operating levers, each part of which is necessary for exact timing. To see clearly the pinion E and the projecting arm D already described, the snail and hour wheel H must be removed. Then we see how the arm D, as it comes round each hour, raises a lifting lever K as the time for striking approaches. This, in turn, raises another lever L which has two projecting pins, M which arrests a little rotating disk, and N which holds up the rack. The disk is part of the striking train, and as it rotates it has a projecting pin which engages with the teeth of the rack.

Once the lever K begins to move we may watch what happens. It goes on moving until the lever L is sufficiently raised to lift the pin M off the disk. This turns a little, and so does the fly at the top of the striking train; but only for an instant, as meanwhile the top of lever K, passing through a slot, stops one of the wheels of the striking train, and it is this which now holds up the train. This happens at about 4 minutes to the hour. By

A star wheel in the striking train raises a pin which works the hammer lever.

3. As L is raised, M releases rotating disk, but the top end of K, passing through a slot, holds up the striking train.

1. Normal position of the levers: the pin M holds the rotating disk and the pin N holds up the rack.

2. With rack, 'snail', and hour wheel removed, we see how, as the hour approaches, the arm D raises the lever K.

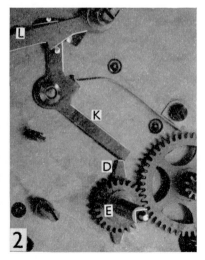

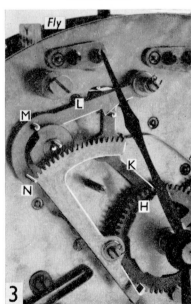

4. *L* rises more until *N* is clear of the rack, so the rack falls and its tail then makes firm contact with the 'snail'.

5. Parts removed to show how, exactly at the hour, *K* slips off *D*, and the upper end of *K* springs back and so releases the striking train.

[Clock by courtesy of The Garrard Engineering and Manufacturing Co. Ltd.]

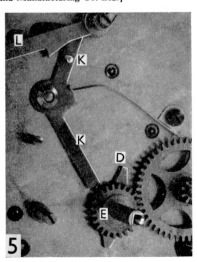

the time it is only about 2 minutes to the hour the lever *L* is sufficiently raised to bring its pin *N* clear of the rack. So the rack falls, and the tail of the rack has plenty of time to settle in the correct position on the snail.

As the hour approaches, the levers are raised to their highest position, and at the exact moment when the minute hand reaches the hour, the lever *K* slips off *D*, and as it springs back its top end, which up to now has been holding the striking train, jumps clear, and the striking begins. The wheels are in motion, the fly whirrs round, and at each blow of the hammer the little rotating disk makes one turn, and in so doing the pin projecting from it moves the rack one tooth. In this way it 'gathers up' the rack one tooth at a time. When striking 6 the snail holds the tail of the rack in the position in which there are 6 teeth of the rack to be moved. When they are all gathered up, the lever *L* falls by its own weight, its pin *N* holds the rack so that it cannot move until next released; and at the same time its pin *M* drops into the notch of the rotating disk, thus arresting the striking train. For half-hour striking, when a single stroke is needed, the rack is not released at all: exactly opposite the arm *D* is a shorter arm which only lifts the levers part of the way—just sufficiently, in fact, to allow the rotating disk to make one turn, without raising the pin *N* far enough to release the rack.

With the older striking clocks it is not possible to turn the hands backwards without doing damage; but modern striking clocks are generally made foolproof, and the arm *D* trips past the lifting lever *K* without harm if someone does push the hands backwards. It is advisable not to turn the hands of a good clock backwards as it sets up an unnecessary strain; the hands should be set forward to the time required, or the clock can be stopped and restarted later. Sometimes provision is made so that a striking clock can be used silent, that is, by not winding the striking train. As the tail of the rack then remains resting on the snail it has to be shaped so that it rides over the snail between the 12 and 1 o'clock positions, or the clock will stop.

In a clock that chimes at each quarter hour there are three springs to wind: the time train is usually in the centre, the striking train on the left, and the chiming on the right. The chimes are controlled by a locking-plate mechanism, and a series of cams operate the hammers. In modern clocks the chimes are self-correcting; that is, even if the chiming period becomes out of step, it is automatically corrected at the next time of operation.

9 *Big Ben,* the Great Clock at Westminster

BIG BEN is the popular name for the large clock in the clock tower of the Houses of Parliament at Westminster, London. Its chimes, relayed by the B.B.C., have become familiar at home and overseas. It may be said to be a direct descendant of the King's clock in the palace of Westminster mentioned on an earlier page. But it is a very notable clock for several other reasons. It was specially designed for the clock tower when the Houses of Parliament were rebuilt, and it embodied features that were then quite new in clock design. It has now been going for over 100 years, and from the start, so successful was its timekeeping that some of its features, particularly the escapement used, have been copied throughout the world for turret clocks—the name generally given to large public clocks. It may be looked upon as the final perfection of the clockwork turret clock, before electrically driven clocks came into use. It keeps such good time that the rate is usually no more than a few tenths of a second per day, and only occasionally does its error exceed a whole second. And it is still one of the largest and most powerful chiming and striking clocks in the world.

There are 4 dials, each $22\frac{1}{2}$ feet diameter, and the minute marks round the outer edge are 1 foot apart. The hands are balanced by counterweights on the driving shafts inside the tower, and also inside are electric lights to illuminate the dials. The belfry is in the upper part of the tower, above the dials; and in the topmost part is a powerful light which is kept on only when the House of Commons is sitting. Though church bells are usually fixed to a wheel and, by means of the bell-ringer's rope, are swung right round to ring, bells that sound the hours are usually fixed, and are tolled with a hammer or clapper. This bell is 9 feet diameter and weighs $13\frac{1}{2}$ tons. It was originally named *Big Ben* after Sir Benjamin Hall, the Chief Commissioner at the time when the buildings were completed; and the name has since come to be used for the clock itself. The note of the great bell is E, and around it are fixed the four smaller bells used for the chime. Above the bells in the belfry is the B.B.C. microphone installation which picks up the sound for broadcasting.

The movement of the clock is housed in a room below the dials. The clock is driven by weights hung on steel ropes which descend through the floor into a weight shaft running down the length of the tower. The clock was originally wound by hand, but it is now wound by an electric motor. This is necessary three times a week, and an automatic mechanism, operated by cams, switches off the motor during the periods when the clock is chiming and striking and when winding is complete. Steel ropes run from the clock movement up to a lever chamber below the belfry, and the levers operate the bell hammers. The pendulum beats intervals of 2 seconds at each swing, and its effective length is a little over 13 feet. The hands of the dial, therefore, move on a little every 2 seconds. The time train of gears in a turret clock has to withstand unusual variations in the load: a high wind, or snow may retard the hands, or a lot of starlings may decide to sit there. As such variation would affect the time-keeping, the force required to keep the pendulum swinging is kept uniform by means of a special escapement. This is designed to lift levers, first on one side and then on the other side of the pendulum rod. As each lever is released it gives a tiny push to the pendulum. So whatever variations there may be in the force operating the escapement, the amount by which the levers are lifted is the same; and as they fall they give an unvarying push to the pendulum, so keeping its swing constant. This, called a gravity escapement, is much used for turret clocks; it has on it a fly or fan not unlike that of a striking train, except that the motion is intermittent.

To avoid stopping the pendulum when the rate of the clock has to be altered, there is a flange on the upper part of the pendulum rod which serves as a tray on which weights in the form of coins may be placed. The addition of a half-penny on this flange, for example, will make the clock go faster by $\frac{1}{5}$th second in a day, or the removal of a halfpenny will make it lose $\frac{1}{5}$th second. The movement was originally constructed by a famous firm of clockmakers, E. Dent & Co., who still, after several generations, have the task of keeping the clock constantly oiled and maintained. Its fine time-keeping is appreciated not only by Londoners but by those all over the world who hear its broadcast chimes.

Big Ben: the great bell and its heavy clapper.   Part of the movement, with pendulum on left.

10   Balance Wheel and Lever Escapements

FOR watches or portable clocks the pendulum is unsuitable, so the most usual time-measuring device is the balance wheel with its balance spring. This type of escapement, being very small and delicate, is usually more expensive, and needs more frequent cleaning and oiling than a pendulum escapement. The balance wheel, instead of turning round and round, swings to and fro. The amount of swing exceeds one revolution but must be less than two revolutions: after which it swings the same amount in the opposite direction. The swing is controlled by the balance spring, a very fine spiral, often called a hair spring.

Variations of temperature affect a balance wheel more than a pendulum. Heat expands the wheel and also makes the spring less stiff, so that the wheel swings more slowly. A ten degree rise in temperature may cause the loss of a minute in a day if nothing is done to compensate for these changes. There are various ways of dealing with the problem, the most usual being to make the rim of the balance wheel of two metals, with the rim cut through in two places, as shown in the diagram. The tiny screws on the rim are placed to get the best effect. The four longer screws at the quarter positions are used to get the wheel poised exactly in balance. Also any opposite pair may be turned an equal amount to adjust for the correct rate; screwing in makes the wheel turn faster, and screwing out, slower. These are fine adjustments performed by the maker. A simpler method is arranged for the use of the owner. The outer end of the balance spring is fastened to a fixed support; but a little before this the spring passes between two pins close together like fingers. The position of the pins is adjustable, and so the effective length, and therefore the stiffness, of the balance spring may easily be altered. The pins are set on an arm pivoted above the balance wheel, and one end of the arm serves as a pointer moving over a scale marked S (or R) for slower, and F (or A) for faster.

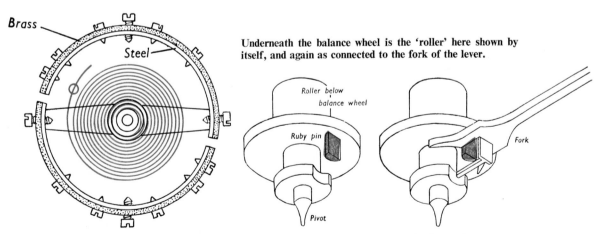

Underneath the balance wheel is the 'roller' here shown by itself, and again as connected to the fork of the lever.

Enlarged view of balance wheel and balance spring. As brass expands more than steel, the free ends curl slightly inwards and so compensate for any expansion of the wheel, and weakening of the spring.

Sometimes a balance spring has its outer end bent over and above the rest of the spring so as to be fixed nearer the central pivot. This, called an overcoil, enables the spring to remain concentric with the spindle as it dilates and contracts.

There are other ingenious ways of ensuring good timekeeping in spite of temperature changes. The simplest is to make the balance wheel all of one piece of nickel–iron alloy, 'invar', and for the balance spring to be of nickel–chromium alloy, 'elinvar'. These metals change hardly at all, within normal ranges of temperature. If it is required to compensate even for very slight differences, that can be achieved by making the two spokes of one metal, and the ring-like circumference of another. In this way, at a departure from normal temperature the wheel becomes slightly elliptical, thus accurately preserving the period of swing.

After so much consideration of the wheel, we should notice that variations in the elasticity of the balance spring are many times more important than any special construction of the wheel.

There are various types of escapement for use with a balance wheel; the most common is the lever escapement. This is shown in the diagrams much enlarged. The mechanism is usually small and partly hidden by the wheel and its spring. The sequence of operations is not unlike that of the anchor escapement described on page 17. The pallets as they stop and release the teeth in turn give the rapid ticking that we hear.

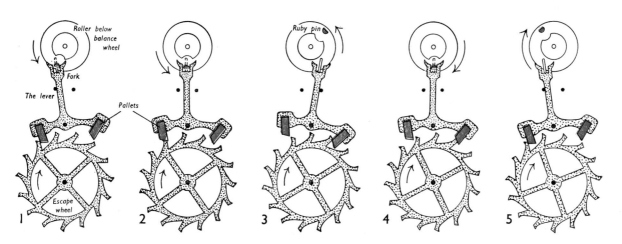

The operation of the lever mechanism. When the lever swings over to the left, the first pallet stops one of the teeth of the escape wheel. 1–On the return swing of the balance wheel, the ruby pin enters the fork. 2–As the lever moves, the escape wheel is released, and in turning, the toe on the tooth gives a tiny push to the pallet. This is transmitted by the fork to the ruby pin, and keeps the balance wheel in movement. 3–As the lever swings over to the right, the second pallet stops a tooth. 4–When released, this tooth in turn pushes the pallet. 5–The lever moving to the left, the first pallet again stops a tooth, and the whole sequence shown is repeated continuously.

11 Watches

NOWADAYS the lever escapement is almost always used in watches. The balance wheel and spring may be compensated for temperature changes in the way already described, or they may be made of the new alloys which are scarcely affected by such changes. Some cheap watches, which have no kind of compensation, work fairly well because when worn on the person they are kept at a reasonably even temperature.

The escapement mechanism is small, delicate, and must be very accurately made. If the mechanism ticks at the rate of 5 times a second, this means 18 000 ticks an hour, or 157 788 000 ticks in a year. With all this incessant movement going on, it is no wonder that a watch needs cleaning and oiling after about a couple of years. Tiny bits of dust and grit get into the mechanism, or lodge in the bearings, where oil, during the course of time, tends to get gummy and stiff. When a watch starts persistently running slower, it is usually a sign that it needs cleaning. Watches and domestic clocks do not have their gears oiled: only the pivots are oiled, and the pallets receive a mere trace of thin oil.

To achieve smooth running and avoid friction, the wheels of, for example, a bicycle are made to run on ball-bearings. But these cannot be fitted to the pivots of an escapement because the mechanism is too small. So the bearings in which the wheels and pivots turn are made of semi-precious stones, generally of a synthetic sapphire. The stones are like tiny disks or buttons, and through the centre of each is a minute hole into which each pivot fits. Some, called 'end stones', have no hole but take the thrust of the pivots of the balance wheel when the watch is turned about in different positions. The pallets are made of stone, as also is the pin, called the ruby pin, on the spindle of the balance wheel. This is called a jewelled movement, though the jewels are usually too small to be obvious. They are used not for ornament but because they are smooth and hard, and so give little friction and long wear.

A watch is best wound at the beginning of each day so as to provide it with plenty of power to withstand the bumps and moving about it gets in the daytime.

Diagrammatic section of the escapement mechanism showing, in red, the position of 11 jewels.

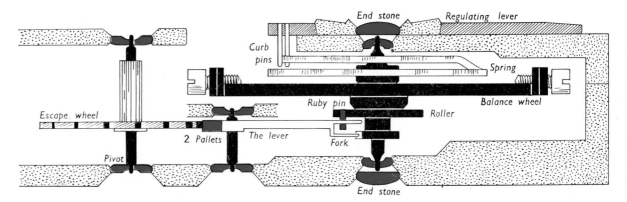

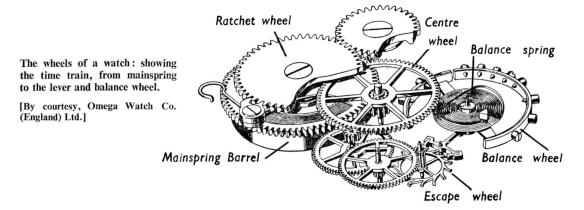

Centre wheel

Balance spring

The wheels of a watch: showing
the time train, from mainspring
to the lever and balance wheel.

[By courtesy, Omega Watch Co.
(England) Ltd.]

Mainspring Barrel

Balance wheel

Escape wheel

On the forward turn of the knob we feel the resistance of the mainspring.
The clicks are from the ratchet pawl holding the spring from running back. On
the reverse turn of the knob the clicks are just from the free-wheel action. The
usual provision for setting the hands is a gear-change device; on pulling out the
winder knob a little way, the winding gear is disconnected and the hand-setting
mechanism is engaged. (The top of page 19 illustrates and describes how this
is done without stopping the works.) If there is a seconds hand, it may be fitted
to the fourth wheel of the time train. More usually it is a centre seconds hand,
driven by additional gearing.

'Automatic' wrist watches have a self-winding mechanism driven from a
pivoted weight that is free to swing round the whole 360° inside the case as
the watch is moved. By means of reducing gearing a swing in either direction
gradually winds the mainspring during the day, more than sufficiently to power
the watch till next day. An automatic watch cannot overwind itself.

The dial of a watch or clock can also be called the face; the pointers we should
always refer to as the hands. The most famous clockmakers whose work is
greatly prized for its fine craftsmanship, always made plain well-proportioned
dials. Roman numerals were the rule, and became traditional over the centuries,
but ordinary (Arabic) numerals are more often used now, or numerals are
omitted and replaced by 'batons', more like the Roman I. This shows we are
able to read the time if we see only the pattern made by the hands. There is
wide variety of design, sometimes so over-decorated that the designers seem to
have forgotten the purpose of the dial and hands. Many watches and alarms
have luminous hands and numerals. The special compound is painted or printed
on, and contains a weak radioactive source such as tritium, which will keep up
a useful amount of luminosity for years.

12 Modern Watches

PERHAPS it is to measure the increased pace of modern life that so many clocks and watches are wanted. There is world-wide competition to meet the demand. We may think of Switzerland as Europe's main producer, but the industry has spread to America, Russia, Japan, and even to China and India. The earliest record of an English watchmaker is about 1580. For centuries after that, there was a clear English supremacy until the Industrial Revolution when England missed the opportunity to adopt systematic large-scale production. With the aid of a remarkable standardization scheme, Switzerland's watch trade prospered and is still one of their major industries. Britain had no large horological firms until two world wars demanded a home supply of precision instruments. After the wars the manufacturers turned to consumer markets and so began a new British horology using highly scientific methods. Europe's largest clock and watch manufacturer is an entirely British firm who made every part of the 17-jewel watch shown here, in their factories in England, Scotland, and Wales. It has a jewelled lever escapement in shockproof bearings and is anti-magnetic. The same movement is used in watches supplied to the Royal Navy, Army, and Air Force.

[By courtesy Smiths Industries Limited.]

13 A Watch of 1830

THIS watch belonged to Thomas Hood, the author's great grandfather; a farmer of Whig Hall, South Otterington near Thirsk, Yorkshire. It has a verge escapement similar to that used in the very earliest clocks as described on page 13. The balance wheel swings only about half a revolution each way. The movement includes a miniature chain fusee which makes the 'change of gear' compensation for varying tension of the mainspring. The movement is hinged inside a silver case with a bezel and glass, all of which is contained in an entirely separate outer or 'pair' case. This was the usual form of watches from about 1650 to 1850. The total weight is 5½ ounces and the thickness is 1 inch. It is no wonder these were called 'turnip' watches. The top plate of the movement is decorated with hand engraving. This is traditional from much earlier watches; those were often extremely ornate. There are many very beautiful examples in museums or owned by collectors.

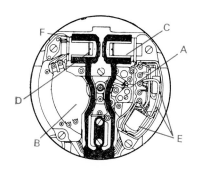

F—Tuning Fork
D C—Driving coils D and C
A—Vibration-counting
    mechanism
E—Transistor, capacitor,
    and resistor
B—Below here is the
    power cell

An Accutron watch.

Accutron circuit.

THIS unusual watch, called the Accutron, keeps more accurate time than a conventional one, and it runs for about a year on one small mercury cell. It has no mainspring or balance wheel. Instead, the time-measuring element is a miniature tuning fork. There is no ticking; only a faint hum if the watch is held to the ear.

The idea of the tuning fork originated as long ago as 1711. To see how it works, a springy metal rod may be clamped at one end in a vice. If we twang the free end, the rod vibrates at a frequency that remains constant if the length remains unchanged. If instead of being fixed at one end, the rod is folded to form a U shape, we have an elementary tuning fork with the advantage that it is now portable. When the fork is vibrating, its two limbs always move in opposite directions.

The vibrating component of the Accutron does not look much like a fork; it is shown in black in the diagram. Its limbs, only 1 inch in length, vibrate 360 times in 1 second. The free ends are cup-shaped and move over fixed electro-magnetic coils energized by the power cell and controlled by an electronic assembly comprising a transistor, capacitor, and resistor; very simple as such circuits go, but without using any moving parts it can rapidly switch on pulses of electricity to the coils precisely as the limbs of the fork are moving outward.

The purpose of the rest of the watch is to count the vibrations and to show the answer on the hands and dial. We might expect that some other electronic process would be required to count the fork's 360-Hz vibrations, but the count is done purely mechanically, by a microscopic ratchet device. Attached to one limb of the fork is a fine, short, flat spring. On the end of this is a rectangular jewel that advances the index wheel a tooth at a time. (There is also a jewelled non-return spring to hold the wheel after each forward movement.) As the index wheel has 300 teeth and is less than a tenth of an inch in diameter, we realize that this is no ordinary mechanism.

Some entirely electronic watches have already been planned and are eventually likely to replace all the traditional designs of watches we use today.

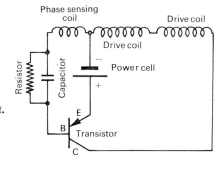

15 The Alarm Clock

EVERYONE knows what an alarm clock looks like. It has two sets of gearing, one the time train, and the other the alarm train, and both must be kept wound each day when the clock is in use. The hands are set by a knob at the rear; another knob turns the pointer on the setting dial for the particular time the alarm is required, and a third knob, either at the rear or at the top of the clock, silences the alarm when it is pressed in. Alarm clocks with a fine quality movement are becoming more popular; but there is also a constant demand for inexpensive alarms. In these, the balance wheel is large and plain and is not compensated for temperature changes. The example below uses a pin-pallet escapement so described because two pins on the lever serve as pallets; these engage with an escape wheel which has plain and widely spaced teeth. In spite of the

[By courtesy, Smiths Industries Limited]

simple construction, the time-keeping is surprisingly good; though with any alarm clock a slight change of rate is no great matter as the clock is reset each night.

The alarm train is very simple, as it is not necessary to have many gears. The mainspring drives an escape wheel with plain teeth, and the two pallets engaging with the teeth are merely parts of the escapement lever, bent over to the right shape. One end of the lever is extended as a long, curved arm to form a hammer, while the other end is extended to engage with the releasing mechanism. As the photograph shows, there is an extra wheel *A* geared to the motion work of the clock, which, like the wheel of the hour hand, rotates once in 12 hours. There is a little projecting lug *B* on it moving against a small disk or cam *C* with a notch in it. This disk turns with the pointer of the setting dial

The front of an alarm clock movement, showing the alarm release mechanism; on right, from the rear, showing alarm escape wheel and hammer.

[By courtesy, Westclox Ltd.]

and may be set for any time in a 12-hour period. When setting, the alarm must be turned only in one direction, shown on the setting dial, or else the projection B may jam in the notch. The flat strip D fixed at one end acts as a spring, pressing B on to C; its other end is bent to hold the alarm escape lever.

As the hours pass, B slides round under C until it reaches the notch; at the time set it jumps into the notch and the strip D, in springing upwards, releases the escape lever which, under the power of the alarm mainspring, dithers to and fro as fast as it can, the end of it acting as a hammer and loudly ringing the bell. To stop it, the silencing knob is pressed and the escape lever is pushed to engage with D once more. This knob must be pulled out again when next setting the alarm. The designs of different makers may vary in detail, but the principle of operation remains much the same.

Stop Watches 16

FOR timing races, for accurately measuring short intervals of time in industry, or for scientific observations, a stop watch is used. Sometimes this looks like an ordinary watch, fitted with a centre-seconds hand which can be stopped and started at will. Pressure on a button starts the hand moving, pressure again stops the hand so that a reading can be taken, and pressure a third time causes the hand to fly back to the zero position, ready for starting again. For most purposes the time of day is not wanted, so more usually the stop watch has a dial graduated only in seconds, with a smaller inner dial to show the minutes.

Since pressure on the outside button sets the movement going, and pressing again stops it, the operator can keep his eye on the race, or whatever it is that he is timing. The seconds hand shows intervals of $\frac{1}{5}$th second, and with balance wheels swinging only five times a second, shorter intervals cannot be measured. But special stop watches are made with balance wheels swinging very fast, 30 times a second or more, and readings of greater accuracy are possible. Special dials are made for various sports and games, such as for football, water-polo, or yachting, as the example illustrated.

Top: stop watch reading to one-fifth second. Centre: special stop watch reading to one-tenth second. The smaller dials show minutes. Below: yachts are given 5 minutes to line up for a race, and this timer is numbered in reverse.

[By courtesy, Smiths Industries Limited]

17 Special Clockwork

MANY ingenious things have been devised for the clock movement to operate. The early striking clocks worked movable figures; we have all seen a cuckoo clock; musical clocks play a tune, the mechanism having a drum with projecting pins, like a musical box. A repeating mechanism was sometimes fitted to striking clocks, so that at any time—in the dark, for example—if a knob is pressed or a string pulled, the clock strikes again the last hour and the last quarter. Clocks have been made with additional small dials to show the equation of time, and to indicate when the mainspring needs rewinding; and some clocks showed the times of high and low tide at a given place. A close approach to perpetual motion is achieved in the Swiss-made Atmos clock. This is automatically driven by the daily changes in the air temperature; it runs year after year without attention.

It is possible to obtain wrist watches which have a calendar dial showing the day of the week and the date. This is done by additional gearing on the motion work; but the days have to be reset at the beginning of each month. Clocks incorporating a perpetual calendar have been made, and these automatically allow for leap years as well as for the different number of days in the different months. These are not as complicated to make as one might think: the necessary mechanism only requires four special wheels and some half dozen levers. It would, therefore, not be difficult to standardize such an addition to the dial, for indeed the clock face is the most logical place on which to look for the day of the week and the date.

A special clock used in factories is the Time Recorder. There are many varieties, but in its most common form it prints the immediate time in hours and minutes, a.m. and p.m., on a card. Every worker has his own card, and as he enters or leaves he places his card in a slot below the dial, presses a lever, and the machine stamps the time. The method is also used for costing, the time being stamped at the beginning and end of any particular job to show how many hours must be charged. The clock movement is connected to a series of wheels which have on their rim the figures in relief. Those for the minutes are moved only at minute intervals; the figures for the hours are moved hourly. Thus at any moment the required figures are in the right position to stamp the time.

Enlarged view of a calendar dial on a modern wrist watch, showing the day of the month.

[By courtesy, Venner Ltd., New Malden]

EARLY this century it was still possible to see the lamplighter going on his customary round as dusk fell, lighting the gas in the street lamps one by one. As electric lighting came into use, one switch might work several lamp standards but it was still necessary to switch on and off by hand. Nowadays, street lamps, advertising signs, and shop windows are automatically controlled by time switches. A few still need winding once a week, but most of them are self-operating and can run for years with very occasional attention. They may be driven by a synchronous motor of the kind shown on p. 37 but as these motors stop if the power is interrupted, the typical time switch shown here is a combination of mainspring drive and motor. The mainspring stores at least nine hours' driving power; the motor runs continuously, geared to drive the spring through a slipping clutch; this avoids over-winding. The illustration is of the complete movement, removed from its watertight case. There is a modern escapement capable of a rate that is accurate within ± 5 minutes each month.

The strongly made dial, with its adjustable segments, makes one anticlockwise rotation each twenty-four hours. Correct time is set by hand. The dial is on a fairly stiff friction-drive spindle so that any hour or part hour can be brought opposite the fixed pointer marked TIME.

The switch levers themselves are hidden below the name-plate. The screw pin A, as the dial slowly approaches dusk time, engages with an ON mechanism, causing two silver contact buttons to spring firmly together. Similarly as dawn approaches, the screw pin D, at a different level, has come round with the dial and it engages an OFF trip, making the contacts snap apart sharply.

(If required, there is a second pair of OFF and ON pins, B and C, to save illumination during the small hours when few folk are about. The ON/OFF disk at the foot is to enable a servicing operator to test the switch by hand.)

33

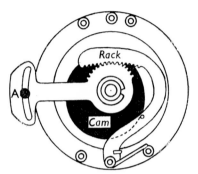

The calendar mechanism under the dial is worked by the star wheel as it trips past the fixed pin. The diagram shows how the arms holding the dial pins are moved by a slow-turning cam (black) which works a rack mechanism.

An ingenious feature of this street lamp switch is that pins *A* and *D* are not fixed directly to the dial but are on segments positioned by a cam, so arranged that adjustment is made each day for the progressive changes in the times of sunset and sunrise. As these times vary not only with the seasons but with the locality, the cam is designed to compensate for the time of year and for the area in which the dial is to be used. The diagram above shows how the segment position is altered by the cam which rotates once a year, driven from a gear train that daily receives a slight advance. This is by means of a five-tooth 'star' gear which engages once a day with a fixed pin in the clock base (see edge-on view). Initial setting of the cam at a correct position is provided for by a disk marked off in months, and seen through an aperture in the main dial.

To control the burners of central heating boilers, rather similar timers are used, but they need no solar mechanism and instead have provision for simple programming of the heating according to the day of the week.

In industry and science, automatic timers now have wide application. These range from easily reset clockwork timers for photographic dark-room use, such as controlling the projection lamp in print-making, to complex installations capable of automatically programming any long intricate process, possibly in use for twenty-four hours a day. Machine tools, plastic moulding machines, and automatic etching machines for printers' blockmaking, can all be automatically timed in this way.

The typical auto-reset timer (p. 35, left) also has a small synchronous motor to drive it. Control can be remote, from a push button or from another mechanism. This timer is available in twelve ranges, giving from 5 seconds full scale reading (half a second per scale division) up to 12 hours full scale (1 hour

A Synchronous timer with a wide range of possible applications.

[Both illustrations by courtesy of Londex Limited, London]

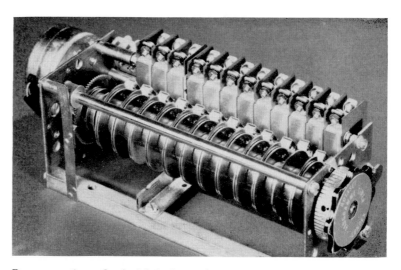

Programme timer. On the left is the synchronous motor, geared to a camshaft on which can be 3, 6, 9, or 12 adjustable cams settable to open or close the contacts at the rear.

per division) and there is a choice of output terminals offering the possibility of elaborate interlocking with other equipment. There is even an 'Emergency Stop' circuit. Timing is normally initiated by a brief pulse of current, originating from near or far. The synchronous motor may either run continually, when a solenoid clutch starts the timing train; or the motor may normally be at rest. There are three stock types of this timer, to meet most requirements of simple timing.

The rather more involved timer shown (right) consists of a synchronous motor driving a gear train which rotates a shaft on which are 3, 6, 9, or 12 cams, each separately adjustable so as to operate any required circuit at a pre-arranged time and for a required period. Time ranges are from one rotation per minute, up to one rotation per week. Such a timer could be used to programme an installation of heaters, solenoids, motors, pumps, and so on, to complete an industrial process without any need for human intervention.

Timers so far described all utilize synchronous motors. But there is a simple electronic device for timing without the use of clockwork, or any motor. Its basis is a capacitor (condenser) that is fully charged to start a timing period. Then by means of adjustable resistors (resistances) the charge is slowly leaked away so that when it reaches zero or almost zero, electronic circuitry signals the end of the measured period. In general these condenser timers are not for high-accuracy timing; their advantage is the absence of moving parts to go wrong, and the ability to stand up to adverse conditions.

19 Electric Mains Clocks

OUR public electricity is 'alternating current'; that is, a flow of electrons that changes its direction 100 times in a second. From a zero point, the flow increases to a maximum; returns to zero; builds up in the other direction to a maximum; then once more returns to zero. The sequence of two alternate flows is called a 'cycle'. Thus the number of complete cycles in one second is 50. This is our national 'frequency'. In U.S.A. it is usually 60 or 25.

We can very easily prove that our current is intermittent if we take a hand mirror and view in it the light from any 'gas-discharge' lamp such as a neon lamp or neon sign, or one of those mercury or sodium street lamps. If the mirror is rocked to and fro, the image of the lamp is broken up into many bright patches separated by spaces of absolute darkness. The lamp goes on and off 100 times a second, because a pulse of current in either direction will light it. (The demonstration can be fairly effective using a fluorescent tube lamp, but fails with an ordinary lamp bulb because its filament keeps hot during the 'off' periods.)

The idea of time-keeping from the mains alternations was thought of before 1900 but only became practicable as larger areas of supply became interlinked and their combined generators could maintain an accurate enough frequency. By 1930 half the power supplies in Britain were time-controlled. Now it is possible anywhere on our mainland to work a clock from the mains and for the clock to continue to indicate, within a few seconds, quite good time. An electric mains clock has no pendulum, balance wheel, or escapement, and needs no winding. It is driven and timed by a small motor of the kind described as 'synchronous' because it moves in exact step with the mains alternations. The hands of the clock are operated through a short train of speed-reducing gears, so that it is quite easy for a small motor to work a large dial.

The clock motor consists of a magnetic 'rotor' which revolves in a fixed electro-magnetic 'stator'. The fine insulated wire winding of the stator produces what is in effect a rotating magnetic field, firmly carrying the rotor with it. Once rotation is started there is no slip; the motor speed continues in exact proportion to that of the mains alternations. These clock motors can either be self-starting (any interruption in the supply causes a 'slow' time indication) or the other kind, needing an initial push-off but with the advantage that the clock shows correct time or is obviously stopped. Earlier mains clocks had some form of indicator disk to show 'clock working'. The same purpose is served by the more general use of a centre seconds hand.

There is a wide choice of mains clocks to suit most purposes, from large public clocks down to bedside alarms. They can be mass produced to sell at such low prices that, as with the clockwork alarm, there comes a stage when it is cheaper to replace than to repair. The simplest mains clock keeps as good time as the most expensive.

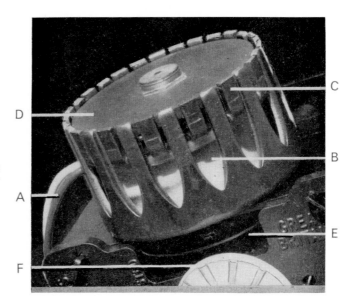

The synchronous motor of a mains clock, shown here much larger than actual size

A—wires from electric mains
B—stator coil
C—stator's 30 pole pieces
D—magnetic rotor disk
E—part of reversing mechanism
F—thumb knob to set hands

How the time gets into the mains is that the Central Electricity Generating Board has a National Control centre whose aim is to make sure there are sufficient generating sources feeding into the supply Grid throughout our mainland to meet the public demand at any hour of the day or night. When a generator is run up to speed and then connected to the Grid, its frequency thereafter locks to exactly the same as that of all the other generators feeding the Grid. On the rare occasions when the Grid has more current taken from it than the available generators can readily maintain, then the frequency falls slightly below 50. The C.E.G.B. is required by law to keep the frequency between 49·5 and 50·5 cycles per second. The National Control is able to work well within these limits because it has available all our coal-fired, atomic, and hydro-electric stations and is continually bringing in or dropping out sources of supply. To keep the frequency absolutely steady is not so far practicable; but losses or gains are very soon rectified, so that a mains clock always shows about the right time. We should think of this when we use the radio pips to set our mains clock at the best hour which is 7 a.m. when mains time is correct to within a few seconds. No further adjustment should be made, even if the clock continues running for years.

The purpose of the C.E.G.B. is to provide electric power. The time feature of the mains is a gift to electricity users, who have found that especially where large dials are wanted mains clocks are the most robust and trouble-free clocks there are.

37

20 Electric Winding

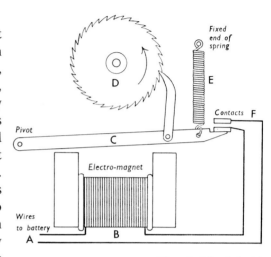

TIME-KEEPING is becoming more and more dependent on electricity as its driving power. As described on page 22 a large motor automatically winds *Big Ben*, a task that used to take five hours' work by three men, three times a week. Most public clocks are wound by electricity. Some of London's Underground stations have for many years used 'self-winding' clocks operated by two circuits; one for the winding; the other to reset the hands every hour by a signal from a master clock.

There are many inventions for driving smaller clocks by electricity. With the improvement of batteries to their present reliability and long life, battery-driven clocks are increasingly popular. The one shown below is made in Germany, runs for 1000 days on one $1\frac{1}{2}$-volt cell (*A* in the illustration), and has a 7-jewel movement in a transparent case. *B* is the electro-magnet; *C* the armature. It works like this—the extended lever from the magnet armature carries an electric contact. Slowly approaching it is another contact attached to the arm of the ratchet pawl. At the instant the two contacts meet at *F*, current flows and the armature contact gives a kick to the pawl contact, impelling the arm rapidly back about a third of a turn, and carrying with it the mainspring barrel *E*, winding the mainspring enough to last about $2\frac{1}{2}$ minutes. After that, the sequence is repeated. The ratchet wheel *D* is designed to be just massive enough to stay where it is while the pawl is moving back in the brief winding stroke. No retaining pawl is needed.

A sharp click is heard at each winding. So this sort of clock is not very suitable for a bedroom or living room. But its plain $5\frac{1}{2}$-inch dial, enamelled on an 8-inch pottery plate gives the clock a most satisfactory appearance.

The principle of electric winding.

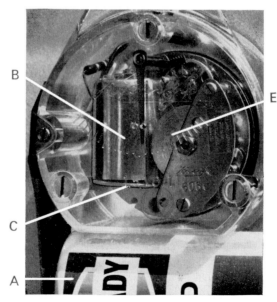

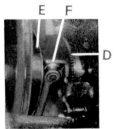

Back view of a battery-driven electrically wound clock, and side view of the same movement to show the contact mechanism.

Detail view of the movement of a battery-driven electronic clock made in Scotland.

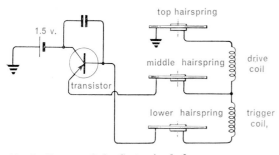

Circuit diagram of the Sectronic clock. Three hairsprings make the electrical connections to the coil system which moves with the balance wheel.

[By courtesy, Smiths Industries]

THERE is another more recent development in battery-driven clocks. Several designs have peculiarly shaped balance wheel assemblies that are really oscillating electric motors. The 'escapements' appear similar to spring-driven ones, but the drive goes the other way. The motive power originates at the balance wheel, and what would be called the escape wheel is now the index wheel, counting the swings and passing on the power it receives to the train of wheels leading to the hands. In the illustration of the clock shown, left, the balance spindle has two wheels on it, each of bizarre appearance but perfectly balanced although carrying part of the magnetic system. The drive coil does not move. With it is a trigger coil that signals a transistor to release a pulse of current to the drive coil. The action is the same whichever direction the wheels are swinging. There is therefore a drive impulse for each half-cycle of the balance.

The first of its kind to use a transistor and moving coil balance assembly is the Sectronic clock. Here the permanent magnet is stationary and the two-part coil moves, attached at right-angles to the rim of the balance wheel. One part of the coil does the triggering; the other receives the battery current impulse as the coil swings through the gap in the magnet. Three hairsprings are used to make electrical connections to the coil. In clocks like these, the transistor is used as a switch that has no moving parts. The demand made on the transistor is only a fraction of its capability; so it should have an extremely long life.

After a while, any such independent clock must have its hands reset. We are reminded here of a very good feature of the mains clocks described earlier. They are self-correcting; or at least the electricity authority continually corrects the mains frequency.

39

22 The Electric Master Clock

TIME-KEEPING in a factory, school, or office block is made simple and reliable by the installation of a master clock system. This comprises a pendulum driven by a battery and operating any required number of 'dials' or secondary clocks, through low-voltage wiring. Such an installation seldom needs attention, and keeps going during a power cut or mains failure.

Let us examine a typical master clock, the Synchronome, now available in advanced form but basically the same as that developed by Frank Hope-Jones in 1905–7. Its operation is illustrated below. The first feature we notice is that this master clock has no train of wheels such as we usually find in a clock. There is a one-second pendulum which is an invar rod that is suspended securely at the top in the usual manner. Near the foot of this rod is a massive cylinder or 'bob' (not shown) also made of solid invar and attached at its exact centre.

So far, we have a complete simple pendulum which, given an initial push, will keep swinging for a long time. To maintain this motion, there is a small bracket about three-quarters of the way up the rod. At the left-hand end of the bracket is a curved profile down which runs a small weighted roller, each half-minute. The curve is so designed that the impulse, brief though it is, begins gradually, builds up to its maximum when the pendulum reaches the middle of its swing, then gradually ends. The diagram shows how the swings are counted. At the last stage, 4, the gravity arm is in its fallen position, where it makes electrical contact. The resultant electrical impulse energizes the electro-magnet which attracts the armature to itself. The gravity arm is thus kicked back to behind its catch, the contact broken, and the armature returned by a spring.

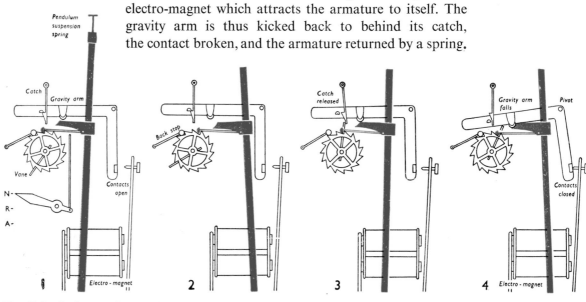

| The light hook on the pendulum rides over one of the teeth on the wheel. | As the pendulum swings to the right, the hook turns the wheel by the amount of one tooth only. | After ½ min. or one turn of the wheel, the vane releases the catch, and gravity arm then falls. | The tiny wheel gives a push to the pendulum bracket, and at the same time the contacts close. |

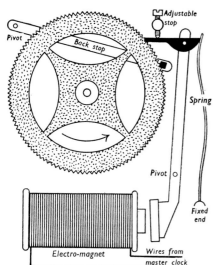

Behind each dial is this driving wheel. Every ½ minute, the electromagnet pulls the pivoted arm. When the pawl (black) is released, it is immediately returned by the spring, and the wheel turns one tooth. The motion is anticlockwise because we look at it from behind the dial.

The same electrical impulse is used to move the pointers, each half minute, of as many dials as are needed. In large installations, dials are grouped to form 'series-parallel' loops, in which they are not operated direct from the master clock but through relays. All the dials of an installation show the same time, and as the standard of time-keeping is high, there is no practical objection to intermittent movement of the minute hands—usually every half minute but in some commercial installations using workers' time recorders, every minute.

For the rare occasions when the indicated time must be reset or corrected, provision is made at the master clock to retard or advance all the dials at once. To indicate what time is being shown throughout the installation it is customary to instal a dial as part of the master clock. The diagram shows how the mechanism of each dial or secondary clock works. It is really nothing more than an adding device, summing the impulses, to show what time they represent. The 120-tooth wheel is fixed to the shaft of the minute hand; the only other wheels are those needed for the hour hand. A master clock may also operate digital clocks which display large numerals, e.g. 23.35 and change instantly, each minute.

The Free-pendulum Clock 23

WHY the Synchronome clock is so successful is that the time-keeping feature, the pendulum, has hardly any work to perform. Once every half minute it unlatches the gravity arm to keep itself going, and pulls the counting wheel round, a tooth at a time. This is the only wheel in the clock.

If the pendulum is relieved of even this amount of work, its natural swing is undisturbed, and the time-keeping is further improved. This is what has been achieved in the Shortt free-pendulum clock, named after its English inventor. The installation consists of two pendulums. One of these, the 'master', has no work to do except to keep the time. The other pendulum is much like the ordinary Synchronome one, but here it is called the 'slave' pendulum, because it does the counting and releases the impulse that maintains the master in motion. Actually, a completely 'free' pendulum is impossible. If it were absolutely free it would come to rest. But Shortt's pendulum achieves almost complete freedom, and is so accurate that for many years it has been used by the world's principal observatories. It is surpassed only by electronic devices such as those described in later pages.

The slave pendulum is regulated to run a tiny fraction slower than the master. A magnetic device ensures that the slave is accelerated at intervals so that the period of the two pendulums never differs by more than a few thousandths of a second. The master pendulum, doing the actual time-keeping, is housed in an air-tight cylinder to reduce air friction. Once the pendulum length has been regulated as in normal clock practice, fine-adjustment regulation is possible by adjusting the small residual amount of air left in the cylinder.

24 The Marine Chronometer

IN the eighteenth century many attempts were made to design a portable clock which would be sufficiently accurate for finding the longitude at sea. This, the marine chronometer, was invented by John Harrison in England, and independently by Le Roy, a French clockmaker. A pendulum cannot be used on board ship, so the chronometer was made like a large spring-driven watch, and an accuracy of about ·3 second per day was achieved. As subsequently perfected by English makers, chronometers have been produced in large numbers from the early nineteenth century to the present day. A ship always carried two or more chronometers, and these were tested and adjusted as soon as it came into port. Nowadays, by radio time signals, the navigating officer may himself keep a constant check on the rate of the chronometer he is using.

A chronometer, being designed for use on board ship, has several special features. It is set in pivoted rings called gimbals so that however much the outer case is tilted by the motion of the ship, the dial and movement remain horizontal, thus avoiding any change of rate which might occur owing to change of position. Most chronometers will run for 2 days, but some are made to go 8 days at one winding. Usually a chronometer is wound each morning by the officer on duty. It is turned face downward as there is a hole at the back into which the winding key fits. In order that the time train shall keep going while winding proceeds, a short bent spring provides temporary power while the mainspring is being wound.

A mainspring, as it runs down, always gets a little weaker; so to keep the power of the drive uniform a device called a fusee is used. This was used in old clocks with the verge escapement, but nowadays is mainly used for accurate clocks and chronometers. The picture shows how the fusee works. When the mainspring is fully wound, the chain unwinds from the small end of the fusee, where it only has a small leverage; but when the spring is almost run down, the chain is unwinding from the large end, and has a larger leverage. In this way the power is kept constant.

A ship often undergoes great changes of temperature; it may pass from the arctic seas to the tropics. The balance wheel, therefore, is specially compensated to minimize any change of rate with extreme temperatures. It is usually made of two metals and has

An 8-day marine chronometer, swung in gimbals, the clamp for which is in the lower corner. The small upper dial shows when the mainspring should be rewound.

[By courtesy, Thomas Mercer Ltd., St. Albans]

42

a pair of round weights set on the rim and arranged to slide for adjustment. The balance spring is fairly powerful, and instead of being a flat spiral it is of helical shape. The escapement is always of a special design, and the diagram shows the principle of operation. Unlike ordinary escapements which release one of the pallets at each swing of the balance wheel, the release action only works at every other swing: there are two swings for each tick, and two ticks per second. This 'chronometer escapement' is particularly accurate, and also has the merit that it needs no oiling.

The chronometer escapement: as the roller turns, the jewel _A_ pushes the flat gold spring _G_ and moves bar _D_ with jewel _B_. The escape wheel turns, and one of its teeth pushes jewel _C_ on the roller, to keep the balance wheel swinging. The escape wheel is checked as _B_ springs back into position. On the return swing of the balance wheel, _A_ trips past _G_ without any action.

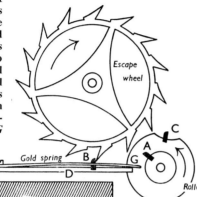

Surveyors, who must be able to find their position more exactly than navigators, also use a similar chronometer, except that, as it is used only for short periods and must be accurately set by radio signals, the large dial shows seconds. When not in use, a lever clamps the balance wheel, and the outer case of the instrument consists of a round metal box with a close-fitting lid like a biscuit tin. Chronometers made for some of the world's most famous ships by the British firm Thomas Mercer Ltd. are fitted with electrical contacts which give $\frac{1}{2}$ minute impulses. These electrical impulses are received by a relay which then transmits a more powerful current for the operation of any number of dials. There is also a self-winding scientific chronometer with similar contacts, which can be used for various purposes; electro-magnets, operated at each $\frac{1}{2}$ minute impulse, turn a ratchet wheel and keep the mainspring always wound.

[By courtesy, Thomas Mercer Ltd.]

On left, the chronometer balance wheel.

[Crown copyright reserved]

A fusee: instead of a cord, the chronometer has a chain, like a miniature bicycle chain.

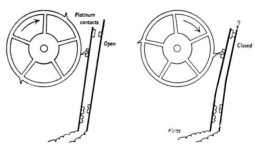

The wheel turns once a minute, and the electrical contacts therefore operate every half minute

25 Time Zones and the Date Line

TIME can be reckoned only for a particular place, and is different for other places not on the same meridian. In a small country like England, the differences between the true time of the eastern and western extremities of the country may be ignored. In a large country like the United States, it is much more of a problem, because for each 15° longitude westwards the time becomes one hour earlier. To avoid the endless confusion of varying times, a system of time zones was adopted in 1883, each zone being like a narrow band running north and south, and the boundaries being adjusted to suit territorial convenience. The time in each zone differs from the next by one hour exactly. Thus, if the time is 11.15 a.m. in New York, which keeps Eastern standard time, it will be 10.15 a.m. in Kansas City, 9.15 a.m. in Denver, and 8.15 a.m. in San Francisco.

This same system has now been extended to cover almost the whole world. The principal areas are shown in this map. The time kept in each zone is that of the meridian which runs down the centre of each zone. As a necessary compromise, the boundaries of zones on land generally follow political frontiers, and the time in some of these zones is not strictly correct for the longitude. But as far as possible, the time in each zone is arranged to be an exact number of hours ahead or behind Greenwich Mean Time; in some countries the difference is to a half-hour, and in a few territories to an odd number of minutes. The largest countries have to be divided into several zones, and the oceans are divided by the lines of longitude shown, giving zones which differ by exactly one hour.

On the opposite side of the earth, at 180° from the Greenwich meridian, is the international Date Line. By a fortunate chance it runs not across land, but over the Pacific Ocean; though the line makes slight deviations from the 180° meridian in order not to divide islands under the same administration. Since the time differs throughout the world according to the longitude, the date line is a necessary convention. Provided we remain in one place we do not need to think about these differences in time; it is only when we move about that they have to be allowed for, and only when we cross the date line that the date has to be changed. Suppose a ship travels eastwards round the world, and correctly puts on the clock one hour on entering each time zone, it will have added 12 hours by the time it reaches the date line. In other words, it will have gained half a day. Now if a ship sails westwards it has to put the clock back, and will have subtracted 12 hours by the time it reaches the date line. In other words, this ship will have lost half a day. This means that each must make some adjustment so as not to get out of step with the calendar. The ship, therefore, that went eastwards repeats a day, and that which sailed westwards misses out one day altogether. Even when travelling a short distance, the same change must be observed whenever the date line is crossed, in order to keep the calendar correct.

44

For example, if a man leaving the Fiji islands and sailing for the Samoa islands arrives at the date line on Saturday night at midnight, he finds, when he crosses the line, that he appears to have arrived on the previous night, for according to the local time it is only the early hours of Saturday, and he has to live all through Saturday again. But suppose, instead, leaving Samoa and sailing to Fiji, he arrives at the date line on Saturday night at midnight: he would then have to drop a day, because it would be Monday morning immediately he crossed the line.

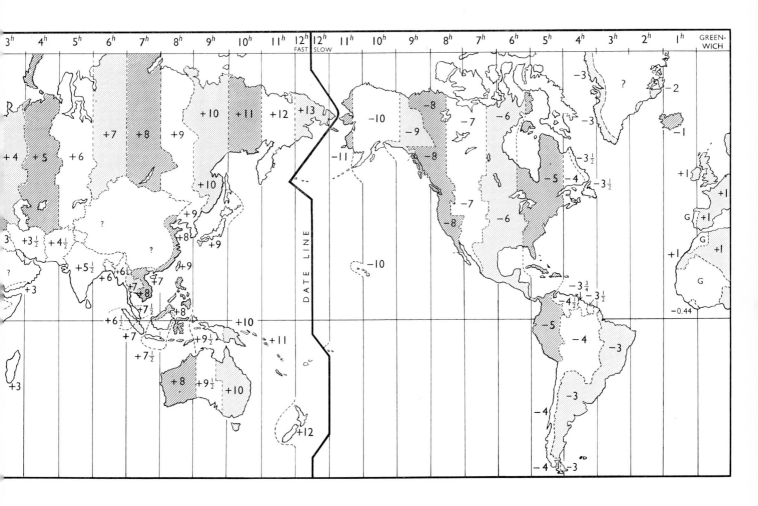

26 The Royal Greenwich Observatory

IN the British Isles, the standard time is one hour ahead of that of the meridian of Greenwich, the north–south line that runs through the old Greenwich Observatory. Charles II founded the observatory in 1675 for the purpose of making tables of the positions of the heavenly bodies to serve as a guide for mariners, and especially to enable them to find their longitude. The original building was designed by Sir Christopher Wren, and numerous additions have been made during its long history. The famous Airy Transit Circle dates from 1851 and is named after the Astronomer Royal, Sir George Airy, who designed and installed this instrument with its large scale divided in degrees for reading the positions of stars. Through the centre of this instrument runs longitude 0°, the prime meridian of the world. In 1854 Airy designed the first reflex zenith tube.

The work of the Time Department is to determine, with the utmost accuracy, Greenwich Mean Time based on the Prime meridian. To do this, clocks must be continuously checked by direct observations of the stars. Radio time signals from Britain and from overseas are regularly monitored, and corrections are published. New forms of transit instruments are now used and, as they are no longer on the prime meridian, allowance is made for the difference in longitude. During the 1939–45 war the Time Service was operated from Abinger, near Leith Hill, Surrey; and from Edinburgh. The task of moving the Royal Greenwich Observatory to Herstmonceux near the Sussex coast was completed by 1958, and it is now re-established there in clearer air, and away from the glare of city lighting. The old buildings at Greenwich have been opened to the public. There is much to see; especially the clocks and astronomical instruments. The original meridian continues to be the basis for world definition of time and longitude.

The Airy Transit Circle at the old Greenwich Observatory.

[Paul Popper]

46

THE method of finding the time by observing the stars is relatively simple in principle. We have first to remember that the position of a star is defined by degrees of declination north or south of the celestial equator (the equivalent of latitude on the earth) and by hours and minutes of right ascension (which corresponds with longitude on the earth). The reason why this is generally given in hours and minutes is that the celestial sphere appears to revolve once in 24 sidereal hours, because of the rotation of the earth. The sidereal time at any place of observation is reckoned from the moment when the first point of Aries crosses the meridian. This is a point on the celestial sphere not marked by a star, but it is the position of the sun at the March equinox. When this point crosses the observer's meridian, the sidereal clock reads 0^h. If a star crosses the meridian $6^h 10^m$ later it is said to have a right ascension of $6^h 10^m$.

In this way, the instant of time at which any given star crosses or transits the meridian is accurately known, and is expressed in sidereal time. If measured

[Photograph specially taken for this book by Photoflight Ltd.]

The old Greenwich Observatory from the air : showing how longitude 0° runs through the building which houses the Airy Transit Circle.

by an ideal sidereal clock, it would be exactly the same each night. In practice, a number of stars are selected for observation, and there is a measurable difference between the time shown by the clock for each transit and the known right ascension of each star. This difference is the measure of the *error* of the clock. By continued observations the *rate* can also be ascertained. (See last lines of page 17.) Once the rate is known, true time can be calculated from indicated time. Time ascertained in this way is sidereal time which can readily be converted to mean solar time for ordinary use.

These star observations are no longer made with a transit instrument but with a Photographic Zenith Tube or PZT which is a special form of telescope camera aimed permanently at the zenith (the point of the sky directly overhead). The diagram shows the PZT's main features. A 10-inch diameter lens collects light from the star and directs it to a pool of mercury from which it is reflected and comes to a focus on a small photographic plate just below the lens. In this position, the plate obstructs about one-tenth of the incoming light from the star, but has no effect on the readings.

The approach of a star from the east causes a faint point of light to cross the stationary plate from west to east. This trail would be recorded as a line. But the plate is mounted in a carriage mechanism which moves it during exposures at the same rate as the image of the star, so a sharp dot is recorded. Four separate exposures are made, each of about 20 seconds: two before and two after

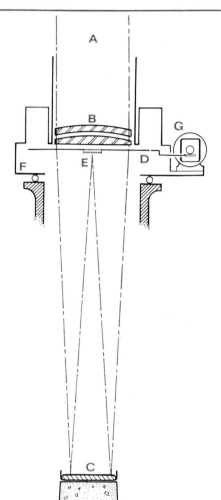

How the PZT works. Light rays, *A*, from the star, are converged by the lens *B*, reflected by the mercury pool *C*, and come to a focus at the stellar plate *E*, which is traversed by phonic motor *G* by plate carriage *D*. The complete rotary *F* turns 180° between exposures.

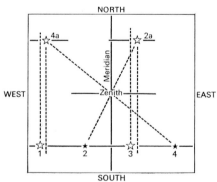

[Drawn from material [by courtesy of the Astronomer Royal]

48

The complete PZT at the Royal Greenwich Observatory.

[By courtesy of the Astronomer Royal]

In this view of the rotary of the Royal Greenwich Observatory's PZT, the plate carriage can be seen in the centre.

[By courtesy of Grubb Parsons & Company Ltd., Newcastle upon Tyne]

836615   D

the meridian transit. This would result in four images 1, 2, 3, and 4 as numbered in the diagram. But after each exposure the whole of the top part of the PZT is automatically rotated 180°. Thus the four images record as the corners of a parallelogram 1, 4a, 2a, 3.

The top of the PZT, called the rotary, only takes about 5 seconds to reverse between exposures. All these motions described are automatic. In addition, attached to the plate carriage are electric contacts which record on punched cards the times when the plate is at certain positions. If the exposures had been at times equally spaced before and after meridian passage, the four dots would form a rectangle, but this will not normally happen, and the relative positions of the dots on the plate must be measured to ascertain the small correction needed to arrive at the exact clock time at which the star was on the meridian.

The observer programmes the PZT beforehand with the star transits he has selected, and can thereafter check the performance of the PZT from a control panel in a nearby building. One photographic plate is used to record all the transits observed in the night. The plate is developed and measured the following day, and the information transferred to punched cards. To obtain the actual readings required, these cards are once a week passed into an electronic computer.

49

28 Quartz Crystal Clocks

FOR many years the time service of the Royal Greenwich Observatory used pendulum clocks. These have now been replaced by extremely accurate electronic clocks. It is the quartz crystal clock that began this completely new approach to the measurement of time. In its early forms, using electronic valves, the clock and its power supply occupied about 20 cubic feet. Advantage has since been taken of the evolution of transistors and new techniques to save power and space. As we see from the illustration the crystal in its 'ovens' and the frequency dividers can now easily be housed in a cylindrical container mounted on the back of a panel 6 inches square.

It was discovered that when the opposite faces of a flat crystal of quartz are given electric charges of opposite sign, the crystal slightly expands or contracts. If then the electrical charges are regularly reversed, the crystal is made to vibrate or oscillate. In the same way that a bell has a natural note or resonance—its audible frequency—so has any piece of quartz a natural frequency. It is a very high frequency far above those of audible sounds. If we now set up our crystal so that it is oscillated by the output from a high-frequency oscillating circuit, and if we adjust the frequency so that it approaches that of the crystal's natural resonance, there comes a point at which the crystal takes control and locks the applied frequency to its own frequency with great precision.

The lens-shaped piece of quartz, cut from a large crystal.

The quartz crystal mounted in its evacuated bulb.

The frequency used in the quartz clocks of Royal Greenwich Observatory is 2 500 000 cycles a second. The total number of cycles *in a day* remains constant from day to day within 2 cycles. This is equivalent in accuracy of time-keeping to about one-millionth of a second per day.

Quartz is a very abundant form of silica SiO_2; it forms the glassy-looking crystals in granite; its grains are numerous in sand, and it is hard and indestructible. In fissures in rocks it sometimes forms huge six-sided crystals. For clock purposes, a piece from a large crystal is cut and carefully polished to the correct shape and size. It is nowadays usually a double-convex lens shape, an inch in diameter, and is mounted inside an evacuated glass bulb encased in two 'ovens', one within the other, so as to maintain a constant temperature not much above normal room temperature.

[By courtesy of Bliley Electric Co., Erie, Pa.]

At the Royal Greenwich Observatory the quartz clocks are installed below ground level, and every care is taken to isolate them from disturbing influences. The clocks are independent of the electric mains. Each cellar has its own separate electricity supply from storage batteries housed on the next storey. These accumulators are duplicated throughout so that one lot may be serviced while the other is in use.

Improving on the pendulum clocks used previously, quartz crystal clocks have the great advantage that they can easily be compared one with another to ascertain which is giving the most consistent performance. Continuous records are made, to extreme accuracy. There is a series of dials located in the control room on the ground floor above the clock cellars. These dials do not indicate the time, but only the difference in rate between the clocks. To record an instantaneous reading, these meter dials are photographed each day. The readings from the previous day are subtracted from them, and in this way the relative gain or loss between clocks is known. Of course, this can only be a test for relative error; the real error must be found by the astronomical observations.

So far we have been talking of clocks, though nothing as yet has indicated the time. The oscillating quartz crystal may be likened to an oscillating pendulum. In a pendulum clock the movement of the escape wheel controls the clock train, which in fact counts up the oscillations of the pendulum and records the time by the movement of the hands of the clock.

To obtain the time from a quartz clock we have to count the pulses of the crystal. How are we to count pulses handed to us at the rate of 2 500 000

A 2·5 MHz Quartz Crystal Oscillator.

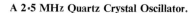

[By courtesy, Sulzer Laboratories Inc., Rockville Md., U.S.A.]

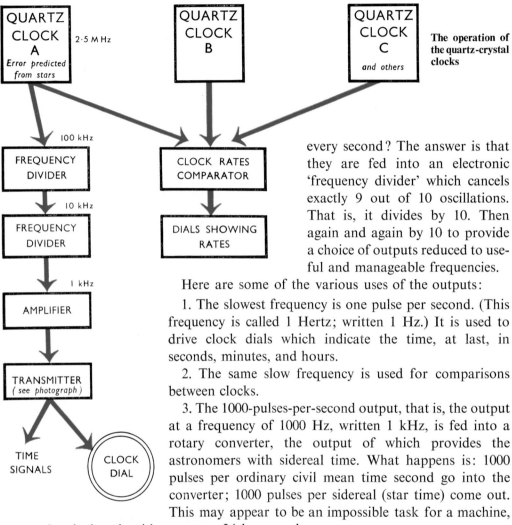

The operation of the quartz-crystal clocks

every second? The answer is that they are fed into an electronic 'frequency divider' which cancels exactly 9 out of 10 oscillations. That is, it divides by 10. Then again and again by 10 to provide a choice of outputs reduced to useful and manageable frequencies.

Here are some of the various uses of the outputs:

1. The slowest frequency is one pulse per second. (This frequency is called 1 Hertz; written 1 Hz.) It is used to drive clock dials which indicate the time, at last, in seconds, minutes, and hours.

2. The same slow frequency is used for comparisons between clocks.

3. The 1000-pulses-per-second output, that is, the output at a frequency of 1000 Hz, written 1 kHz, is fed into a rotary converter, the output of which provides the astronomers with sidereal time. What happens is: 1000 pulses per ordinary civil mean time second go into the converter; 1000 pulses per sidereal (star time) come out. This may appear to be an impossible task for a machine, but it does it with accuracy, 24 hours a day.

4. The 1000-pulses-per-sidereal-second frequency is applied by the astronomers to drive the timing mechanism of the camera of the Photographic Zenith Tube.

How does one 'regulate' a crystal clock? This cannot be done by altering the size of the crystal. That is fixed as nearly as can be to respond to the required 2·5 MHz (megahertz). Whatever the natural frequency of the crystal proves to be, that frequency cannot be altered either. So regulation is applied at the lower frequency stage of 1000 pulses per ordinary second (1 kHz) by feeding this frequency into another special converter which can be adjusted so that it slows down or speeds up the *outgoing* frequency by exactly the amount of difference required. The operator may 'regulate' his crystal clock almost as easily as we regulate our clocks or watches.

[Paul Popper]

The original time-ball at Greenwich Observatory.

THE first public time signal was the ball, 5 feet in diameter, on a mast above one of the turrets of the old Greenwich Observatory, which since 1883 has been raised daily and dropped at precisely 1.0 p.m. At 5 minutes before the hour the ball is hoisted half-way up, as shown in the picture, and to the top at 2 minutes to the hour. As the observatory is on a hill, the time-ball could easily be seen by ships in the London docks, and by it they were able to set their chronometers. After the invention of the telegraph, time-balls were installed in other places; and by 1865 time signals were sent each hour to a telegraph company for distribution over the railway system. Subsequently the Post Office took over the telegraphs and distributed the hourly signals to its branches.

At the castle overlooking Edinburgh, a cannon has been fired each day at 1.0 p.m. for many years.

Everyone is familiar with the six pips of the B.B.C. time signals. These are provided by the R.G.O. from Herstmonceux and are each $\frac{1}{10}$th of a second in duration, spaced so that they indicate the last 5 seconds of the minute. Their accuracy is to within $\frac{1}{20}$th of a second, which is good enough for domestic or commercial purposes.

It is a very different story when we come to the international time signals which are continually available to scientists and electronic engineers. These provide an accuracy that is better than one-millionth of a second.

To distribute the time signals in the various forms in which they are required, transmitting instruments are used in duplicate. This is so as to have one of them continually standing by to take over in case of a fault. Each transmitter is a clockwork mechanism operating a variety of electrical contacts, and is driven by a synchronous motor which has the special name of 'phonic motor' when, as here, it is powered by amplified alternating current obtained from one of the quartz clock frequencies.

The radio carrier frequency—the radiated wave that carries the time signals—is controlled from the same quartz clock source as supplies the time, so that electronic engineers can use it to calibrate—mark off—their dials indicating frequency. The transmissions are thus frequency standards as well as time standards.

One of the duties of the Time Department of the R.G.O. is the checking of time signals; regular comparisons are made with other observatories. We have seen how the time must in the first place be observed by the stars. For making simultaneous comparisons between clocks an electronic device called a decimal counter is used. Overleaf we show one of these in the control room. As with a telephone switchboard, the required connections may be selected according

53

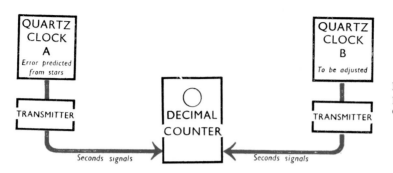

How two clocks are compared by the use of the decimal counter

to the test required. The incoming signal from one clock starts the counter; the signal from another clock stops it. The time difference is then displayed in decimals, to an accuracy that can be read to millionths of a second.

Alongside the decimal counter are radio receivers which are used to check the time signals radiated from Britain, or coming from anywhere in the world. Such signals may be compared with any selected clock.

Many nations now co-operate, and synchronize their time signals as far as possible. Our own station, near Rugby, is at the very centre of England, and from its long-wave transmitter GBR on 16 kHz and its associated high frequencies it radiates signals which are fully described in Whitaker's Almanac. The *American Radio Amateurs' Handbook* (1967, p. 588) describes and shows diagrams of the time and frequency signals available twenty-four hours a day in the U.S.A. and in Canada.

One of the transmitters, with various selector switches and contacts, including the 6 pips for the B.B.C. At the side is the knob for fine adjustment.

The picture on the left shows the switchboard of one of the decimal counters.

[Royal Greenwich Observatory]

THE time-keeping qualities of a quartz-crystal clock depend upon the extent to which the natural resonance frequency of the crystal itself remains constant. In practice, the crystal does slowly 'age'; its frequency becomes higher. This gradual drift in rate may be further complicated by occasional more sudden changes. Attention has therefore been directed to the possibility of controlling a clock, not by a crystal, but by a natural unchanging frequency associated with the structure of an atom or a molecule.

In early attempts to achieve this, a radio frequency was passed through a wave-guide containing ammonia gas. At one particular frequency, interaction between the radio frequency and the ammonia molecules caused absorption, which in turn resulted in a reduced level of output at the receiving end of the wave-guide. Electronic circuits were then used to adjust the radio frequency to give maximum absorption. In this way, the radio frequency was tied to the natural frequency of the ammonia molecule, and was used after suitable division to operate timing circuits and to run a clock.

The most popular form of atomic clock now in use—there are two at the Royal Greenwich Observatory—employs a caesium beam tube. Caesium is a highly reactive silvery-white metal, resembling sodium. The atom of caesium may be pictured as a central spinning nucleus surrounded by electrons, the outer one of which may spin in the same direction as the nucleus or in the opposite direction. According to the direction of spin, the energy of the atom is slightly different. If by applying a radio frequency we change atoms from one state to another, the transition is accompanied by absorption or emission of radiation at a frequency in the region of 9000 MHz. In the two possible states, the caesium atom reacts differently in a magnetic field. So if a beam of caesium atoms is directed between the poles of a magnet, the beam is split into two, and atoms in the two different states are separated.

The principle of the caesium 'clock' is shown here with red and black dots representing caesium atoms in their two different states. The first magnet deflects the black atoms into the tube where some are changed to red by absorption of energy from the radio frequency field. The second magnet directs these red atoms to the detector. The radio frequency is adjusted to obtain maximum

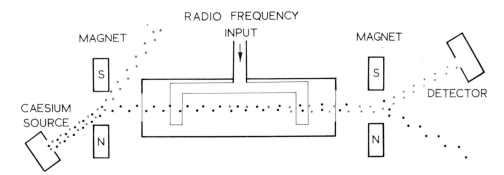

RADIO FREQUENCY
INPUT
MAGNET   MAGNET
S   S
CAESIUM
SOURCE   DETECTOR
N   N

response at the detector. The radio frequency is thus brought into agreement with the frequency defined by the caesium atoms. Comparisons between caesium clocks and Ephemeris Time, E.T. (see next page), established this frequency as 9 192 631 770 cycles per second and caesium clocks are widely used to make Ephemeris Time available. Research and development work continues not only with beam tubes but with other electronic devices known as

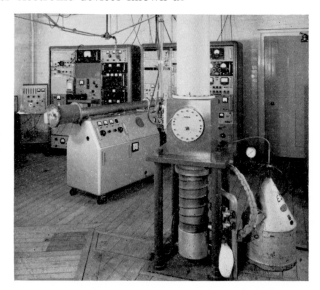

'masers', which use an atomic frequency to generate a radio frequency. In the laboratory a hydrogen maser has already attained an accuracy that is nearly a hundred times superior to that of the caesium beam tube.

As instrumental accuracy increases, so the problems increase: by the theory of relativity it is predicted that the frequency of an atomic standard will not remain constant but will suffer very small variations. In spite of these, all atomic standards used on the earth's surface should be in general agreement; so much so, in fact, that the relationship between Ephemeris Time and the rate of the caesium clock has been turned inside out. The reasons for this are discussed in the next section.

In the foreground is the atomic clock (large caesium beam frequency standard). Behind this and to the left of the photograph is the experimental atomic clock. In the background are the racks containing measurement apparatus and control equipment.

[Crown copyright, National Physical Laboratory, London]

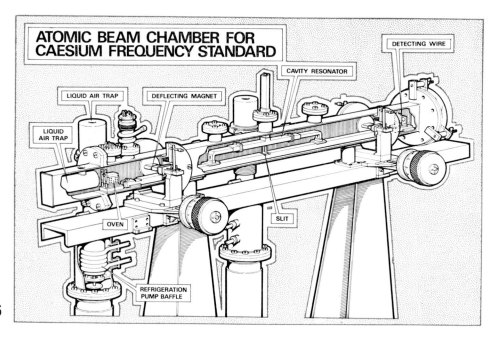

ATOMIC BEAM CHAMBER FOR CAESIUM FREQUENCY STANDARD

DETECTING WIRE

CAVITY RESONATOR

LIQUID AIR TRAP

DEFLECTING MAGNET

LIQUID AIR TRAP

OVEN

SLIT

REFRIGERATION PUMP BAFFLE

[Crown copyright, Science Museum, London]

QUARTZ clocks and atomic clocks maintain such a steady rate that it is now possible to study the small periodic variations which occur in the time it takes the earth to make one complete rotation as checked by the stars. The rate of rotation of the earth has a seasonal variation, due mainly to world-wide weather conditions. The earth turns slightly faster in summer; slightly slower in autumn and winter. In addition, the earth wobbles on its axis. The north and south poles move within a radius of about 30 feet from their average positions. The observations are corrected to compensate for these variations which would otherwise vary the length of the day by about a millisecond.

There are other slight unpredictable variations. The total effect of these can be disregarded for purposes such as navigation or surveying; but scientists have to work to finer limits. The problem was that the second was defined as a simple fraction (1/86 400) of the mean solar day (see page 8). And as this day varied, there was difficulty in describing just what a second was.

A new definition was adopted in 1956. It was based on the time taken by the earth to make one orbit round the sun, i.e. the year. The second was defined as 1/31 556 925·974 7 of a particular tropical year. The new time scale was called Ephemeris Time and the new second the ephemeris second. They were named after a table of figures, the Ephemeris, which astronomers and mathematicians compiled to show the positions in the sky of the sun, moon, and planets at any instant. Clocks were checked by observations of the position of the moon and comparison of these with the table.

This was, of course, a laborious business. Accordingly, in 1967 the second was redefined yet again. This time the definition was made independent of any motion of the earth. It was based instead on the frequency of the radiation emitted by the caesium atom in undergoing a particular change, the same change in fact as takes place in the caesium beam tube (page 55). This second (the third second, or atomic second) is defined as the duration of 9 192 631 770 oscillations of the radiation.

As well as being more accurate than the ephemeris second, the atomic second allows direct comparison to be made with clocks anywhere in the world. This is because the change, and its corresponding radiation, is identical in all caesium atoms everywhere.

57

32 The Telephone *Speaking Clock*

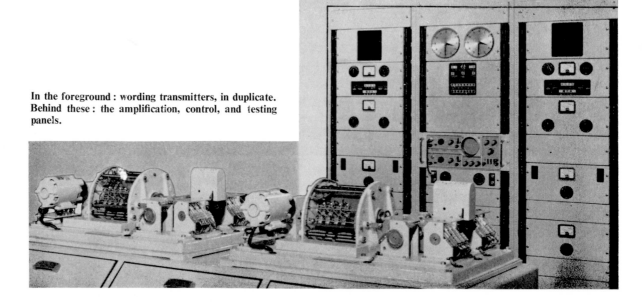

In the foreground : wording transmitters, in duplicate. Behind these : the amplification, control, and testing panels.

LONDON telephone subscribers used to ask their operators the time by the exchange clocks, 26 000 times a week. Profiting by the experiences of some continental countries, the British Post Office Research Department designed an automatic speaking clock and put it into service in 1936 with the dialling code word TIM. Users throughout the country now make about four million calls to the clock every week. In an age when mains clocks and battery clocks are so good and when we can check with the radio broadcast 'six pips' or with Big Ben; or with time shown on the TV screen, it seems odd that there should be any need at all for a speaking clock. Doubtless TIM is so popular because at any time of day or night for the cost of a local call we can dial the clock's code—obtainable from the directory or by asking the operator—and instantly we are connected to a recorded voice giving us the correct time every ten seconds, in the form: 'At the third stroke it will be nine twenty one and ten seconds: pip pip pip'.

The pips are spaced at 1 second intervals, and are of the high-pitched audible frequency of 1 kHz and sound similar to the broadcast pips. The 1936 TIM was kept accurate to plus or minus one-tenth of a second by means of a free pendulum. The whole equipment ran extremely reliably and would have continued to do so for many more years, but to take advantage of scientific and technical evolution, new and completely redesigned installations were made and put into use in 1963. A quartz clock has replaced the pendulum; the earlier photo-recordings on glass disks are

[Illustrations by courtesy of H.M. Postmaster General]

now magnetic recordings on a cylinder; transistors have replaced valves; the speech recording is of better quality; and perhaps best of all, the equipment needs only occasional checking and servicing.

The quartz clock stabilizes the frequency of the current fed to a 50-cycle synchronous motor (small illustration). This motor drives a reducing gearbox to revolve the drum at a speed of 30 revolutions per minute. The drum surface is coated with a thick layer of recording material—a smooth blend of synthetic rubber and magnetic oxide of iron. Spaced along the drum surface seventy-nine separate phrases or words are magnetically recorded, each in its own circular— not spiral—track. Lightly contacting the drum are twelve reading heads; some stationary, others shifted by clockwork to select the tracks required.

  1 track: AT THE THIRD STROKE
12 tracks: IT WILL BE ONE   to   IT WILL BE TWELVE
60 tracks: O'CLOCK   and   ONE   to   FIFTY NINE
  6 tracks: AND TEN SECONDS   to   AND FIFTY SECONDS   and   PRECISELY

The purpose of the rest of the machine is to select the right recordings from the drum and to assemble these at the correct time to transmit the complete announcement each ten seconds.

To ensure that TIM service continues in spite of a mishap or fault, two identical announcing machines are run side by side but entirely independently of each other. One pair of machines is in London; another pair in Liverpool.

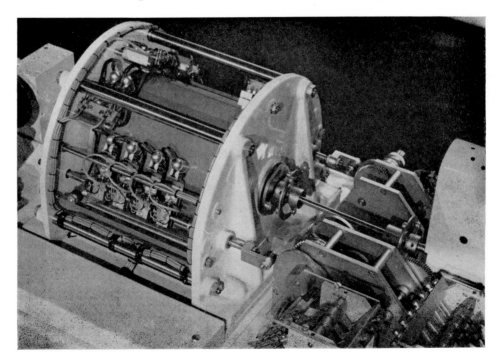

Each installation is normally serving half the country, but if both outputs from one installation fail, the distant centre automatically takes over the entire service. Monitoring equipment seen at the back of the top illustration has two dial clocks with seconds hands. Each clock shows the time represented by one of the stabilized a.c. supplies that drive the drum motors. There are loudspeakers with which to monitor the outgoing announcements, and in the centre is a two-channel oscilloscope to provide visible comparison between the broadcast six pips and the outgoing pips from either announcing machine. You can prove for yourself how closely these coincide if you turn on your radio a few minutes before a 'six pips' hour, and at the same time telephone to the speaking clock.

33 The Progress of Accuracy in Time-keeping

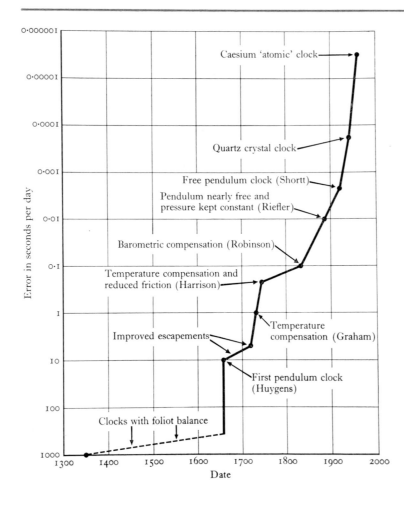

Time-keeping in 1350 was to the nearest quarter of an hour. Progress is now so rapid that even this graph does not bring us quite up to date. Time-keeping has reached an accuracy well beyond the millionth part of 1 second per day. Note what a great improvement the pendulum made, about 1660. The graph's left-hand scale is geometric; each upward step improves the accuracy by a factor of 10.

[Crown copyright, Science Museum, London]

IF we are doing something pleasant such as swimming or playing a game, an hour passes in what seems like a few minutes; but if we are ill or in trouble, the hands of the clock seem to drag round too slowly. Early in history, people were apt to think of time as a series of disjointed events such as successive harvests or notable battles. It was only gradually, as mechanical clocks came into use, that people began to think of time as something flowing continuously and uniformly. To us this seems the obvious explanation; and yet time has no ultimate or absolute measure: it depends on the motion of the observer.

If we look up at the stars on a clear night, we fancy we see them all at a given instant of time. But the stars are very far away, and, according to the time-scale we use on earth, it takes several years for the light from the nearest of them to reach us. For those further off it may take thousands of years for the light to come; and the light we see from the distant nebulae left them millions of years ago. So when we look into space, we are also looking into time, and seeing things as they were at various different times in the past. According to the theory of relativity, time has *no* fixed rate of flow, and may appear differently when measured by different observers. For example, the apparent rate of flow depends on the speed of the observer relative to his measuring instrument. The greater the speed is, the more slowly will time appear to advance. If it were possible to travel at nearly the speed of light, we could circle the distant stars, and on returning to earth find that whereas for us time had only advanced a few months because of our great speed, on the earth itself a million years had passed. This seems odd.

How time has seemed to two famous men is shown in these quotations:

An instant of time, without duration, is an imaginative logical construction.

A. N. Whitehead, *c*. 1940

. . . the direct interpretation of space and time by means of measuring rods and clocks, now breaks down . . .

Albert Einstein, 1935

Books for Further Reading

If you would like more detailed information about time-keeping, here are a few titles to look for in the book shops or in your library. The first two, obtainable from Her Majesty's Stationery Office or at the Science Museum, South Kensington, London S.W. 7, are particularly recommended. Both are by Dr. F. A. B. Ward of the Museum.

1. *Time Measurement*; a historical review (5*s*. 6*d*.).

2. *Time Measurement*; a catalogue of the Museum's collection (13*s*. 6*d*.).

Sundials, How to Know, Use and Make Them; by R. N. and M. L. Mayall (C. T. Branford Co., Boston, U.S.A.).

Sundials Old and New; by Sir Alan P. Herbert (Methuen) (63*s*.).

Britten's Old Clocks and Watches. Revised by Baillie, Clutton, and Ilbert has been the standard work for over 60 years, on horological history (Spon) (7 gns.).

Thomas Tompion, his Life and Work; by R. W. Symonds (Batsford) is a magnificent book about the most famous of all clockmakers.

Country Life Book of Clocks; by Edward Joy (Country Life) (30*s*) is a recent book, mainly for the collector or user of old clocks.

Clocks and Watches; by G. H. Baillie (Associated Trade Publications Ltd., 258 Grays Inn Road, London W.C. 1) (43*s*. 6*d*). This covers time-keeping up to 1800.

More practical books, mainly for watch or clock repairers:

The Mechanism of the Watch; by Sir James Swinburne (A.T.P. Ltd.) (11*s*. 6*d*).

Practical Watch Repairing; by D. de Carle (A.T.P. Ltd.) (32*s*.).

Practical Clock Repairing; by D. de Carle (A.T.P. Ltd.) (32*s*.).

Watch and Clock Encyclopaedia; by D. de Carle (A.T.P. Ltd.) (52*s*.).

Watches; Adjustment and Repair; by F. J. Camm (G. Newnes Ltd.).

For Trade news and progress of invention:

Retail Jeweller, fortnightly magazine, has now incorporated the *Horological Journal* (A.T.P. Ltd.).

Books on more specialized aspects of time-keeping:

British Time; by D. de Carle (Crosby Lockwood & Son).

The Marine Chronometer; by Lt. Comm. R. T. Gould (J. D. Potter Ltd.). The standard work but is only likely to be found in libraries or second-hand bookshops.

Time Zone Chart; is a large wall chart in great detail and has complete information up to date, about the Zones. Published by the Admiralty and obtainable (10*s*.) from J. D. Potter Ltd., 145 Minories, London E.C. 3.

Whitaker's Almanac; each year this includes much about Time and the Calendar, and astronomical facts. Also includes Atomic Time and Ephemeris Time.

The Photographic Zenith Tube and a combined lunar instrument is fully described and illustrated in an article by W. Markowitz reprinted by the University of Chicago Press. It gives the history of the PZT and outlines its present-day use.

'TIM' the Speaking Clock; a complete description of this appears in the April 1963 issue of the *Post Office Engineers' Journal*.

A greatly enlarged print from an actual PZT plate. Note that each star image appears four times.

[By courtesy of U.S. Naval Observatory, Washington D.C.]

© OXFORD UNIVERSITY PRESS 1969

FIRST EDITION PUBLISHED IN 1955
IN THE *Oxford Visual Series*
SECOND EDITION 1969

PRINTED IN GREAT BRITAIN
AT THE UNIVERSITY PRESS, OXFORD
BY VIVIAN RIDLER
PRINTER TO THE UNIVERSITY